# ANCHORING SYSTEMS

# ANCHORING SYSTEMS

*Edited by*

## M. E. McCORMICK

*US Naval Academy, Maryland, USA*

## PERGAMON PRESS

OXFORD · NEW YORK · TORONTO · SYDNEY · PARIS · FRANKFURT

| | |
|---|---|
| U.K. | Pergamon Press Ltd., Headington Hill Hall, Oxford OX3 0BW, England |
| U.S.A. | Pergamon Press Inc., Maxwell House, Fairview Park, Elmsford, New York 10523, U.S.A. |
| CANADA | Pergamon of Canada, Suite 104, 150 Consumers Road, Willowdale, Ontario M2J 1P9, Canada |
| AUSTRALIA | Pergamon Press (Aust.) Pty. Ltd., P.O. Box 544, Potts Point, N.S.W. 2011, Australia |
| FRANCE | Pergamon Press SARL, 24 rue des Ecoles, 75240 Paris, Cedex 05, France |
| FEDERAL REPUBLIC OF GERMANY | Pergamon Press GmbH, 6242 Kronberg-Taunus, Pferdstrasse 1, Federal Republic of Germany |

First edition 1979

**British Library Cataloguing in Publication Data**

Anchoring systems.
1. Anchorage (Structural engineering)
2. Ocean thermal power plants
I. McCormick, Michael E  II. 'Ocean engineering'
621.312'13      TC1650      79-40220
ISBN 0-08-022694-9

Originally published as a special issue of the Journal *Ocean Engineering*, Volume 6, Nos. 1/2 and supplied to subscribers as part of their subscription.

*Printed in Great Britain by A. Wheaton & Co. Ltd., Exeter*

# CONTENTS

*Ocean Engng.* Vol. 6, p. 1. Pergamon Press 1979. Printed in Great Britain

# PREFACE

This issue of OCEAN ENGINEERING contains three excellent works on anchoring. The first paper, "Uplift-Resisting Anchors," is an excellent survey of the field and can be used as a handbook. The second paper, "Preliminary Selection of Anchor Systems for Ocean Thermal Energy Conversion", outlines the thought process in selecting an anchoring system with particular application to OTEC. The third paper, "Single Anchor Holding Capacities for Ocean Thermal Energy Conversion (OTEC) in Typical Deep Sea Sediments", continues the design process into the geotechnical area.

The staff of the U.S. Navy's Civil Engineering Laboratory at Port Hueneme, California, have performed an excellent service to the Ocean Engineering community.

*Michael E. McCormick*
*Editor-in-Chief*

*Ocean Engng.* Vol. 6, pp. 3-137. Pergamon Press 1979. Printed in Great Britain

# UPLIFT-RESISTING ANCHORS

R. J. Taylor, D. Jones and R. M. Beard

Deep Ocean Technology, Civil Engineering Laboratory, Naval Construction Battalion Center, Port Hueneme, California 93043, U.S.A.

## 1. INTRODUCTION

### 1.1. *Purpose of article*

The purpose of this article is to (1) identify and document the status of special types of anchors having the capability to resist uplift forces; (2) provide data on the properties and performance of these special anchors; (3) consolidate the data in order to facilitate anchor selection; and (4) establish a reference that can be readily updated to incorporate new data and new developments. Descriptions and data on anchors that are currently either shelf items or in an advanced stage of development are presented. Also, information on other less advanced designs and concepts is given. Sizes, weights, and operational characteristics of these special anchors, plus methods for estimating their penetration into seafloor sediments and their pull-out resistance, are provided.

This article includes material and information that was possible to obtain within a specified time frame. The development of embedment anchors continues, and additional information will be incorporated as it becomes available.

### 1.2. *Background of uplift-resisting anchors*

As ocean operations and construction have expanded and moved to deeper waters, the need for more sophisticated anchoring systems has emerged. A particular need is for anchors that can resist uplift and are highly efficient, reliable, and light weight where practicable. Other qualities desired are simplicity in handling and the facility for rapid installation.

Anchors that can resist uplift can significantly reduce the scopes of line associated with conventional drag anchors and also the quantity and sizes of accessories. They minimize the need for multileg arrangements to limit watch circle and lessen load-handling equipment requirements. They typically can be installed directly into the seafloor without the necessity of dragging, thus simplifying installation and improving positioning accuracy. They can sustain lateral as well as uplift loading. They broaden the range of feasible anchoring sites, such as on sloping and rocky seafloors, that are considered to be off limits with conventional anchors. They potentially can significantly reduce lowering and placement times, thus making ocean operations less vulnerable to adverse sea and weather conditions.

In deep water, cost efficiency can become the primary reason for utilizing anchors that resist uplift because installation time and line scope become increasingly significant factors as water depth increases. In shallow water, particularly in well-used harbors, uplift-resisting anchors have the advantage of eliminating considerable bottom gear that can be damaged by ship anchors.

Until recently, only a limited selection of anchor types, which were comprised mainly of conventional drag anchors, deadweight anchors, and piles, were available when designing an anchoring system to resist uplift loading. Conventional drag anchors are inefficient for this mode of loading, because they rely principally on their own weight plus that of the sinkers which ensures lateral loading on the drag anchor. Deadweight anchors are heavy to transport and handle for the effective holding to be gained. They are susceptible to drifting, and they are unreliable on sloping seafloors. Piles are limited presently to relatively shallow water.

Commencing in the 1960s numerous anchor concepts were proposed that could counter uplift loading. They included a variety of types, such as propellant-actuated, vibrated, screw-in, implosive, pulse-jet, jetted, and hydrostatic. Some advanced to the development stage, encountered problems, and were abandoned. Others have demonstrated potential workability, but require additional validation testing. A few have been developed to the point of being considered operational hardware.

Despite the progress that has been made with the new anchor concepts, some difficulties remain. Seafloors with anomalous conditions—such as shallow sediment over rock; weathered and fractured rocks; seafloors with gravel and boulders interspersed; and seafloors layered by turbidities—make penetration of the seafloor uncertain and the prediction of holding capacity unreliable. Where seafloor slopes are greater than 10°, the orientation of special anchors for proper penetration is difficult and uncertain.

Deep-water techniques for anchoring in rock are limited to drilled-in piles. Less expensive, more controllable and rapid procedures are needed. Still more expedient means to install all deep-ocean anchor systems are needed. Future installations will impose even more severe anchor requirements. Anchoring systems with 100–1,000-kips holding capacities are envisioned in deep water. Multiple or modular anchors and piles are a potential solution, but knowledge of their interaction and resulting performance must be gained for them to become practical. Anchoring technology is being advanced to meet these challenges.

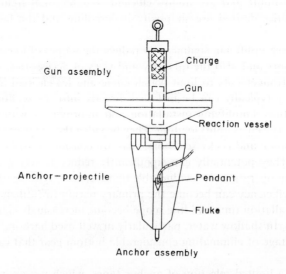

FIG. 1.   General configuration of a propellant-actuated anchor.

## 2. OPERATIVE TYPES

Anchors designed to resist uplift are separated into the following categories:

Propellant-actuated direct-embedment anchors.

Vibrated direct-embedment anchors.

Screw-in anchors.

Driven anchors.

Drilled anchors.

Deadweight anchors.

Free-fall anchors.

Each category of anchor is described, and distinguishing characteristics are identified. Modes of operating, handling, and placing the anchor are given. Advantages and disadvantages are listed. Also, a brief history and the current status of the anchor are summarized. Information on operative designs is given in Section 3.

### 2.1. *Propellant-actuated direct-embedment anchors*

2.1.1. *Description.* A propellant-actuated anchor (often referred to as an explosive anchor) is one that is propelled directly into the seafloor at a high velocity by a gun. Basically, it consists of an anchor-projectile and a gun assembly comprised of a gun and a reaction vessel. Though a variety of forms has evolved, Fig. 1 illustrates the general design of such anchors. The anchor-projectile includes a piston and fluke. The gun incorporates a safe-and-arm device that is actuated by hydrostatic pressure, which arms the gun only after a predetermined depth is attained. A propellant charge, contained in a cartridge, generates the gas pressures that accelerate the anchor-projectile into the seafloor. Whenever possible the gun assembly is recovered and used again. Recovery becomes increasingly difficult at depths greater than 1,000 ft.

There are presently two types of projectiles* for use in sediments. In the first type, the portion which engages the soil to resist pull-out (the fluke) is a rotating plate assembly. It can be either a single-plate construction or a trihedron construction of three flat plates (Y fluke). The plates enter the seabed edgewise. After emplacement, an upward pull on the anchor line, which is transmitted to the fluke at an eccentric connection, "keys" the fluke; that is, the fluke rotates to a position in which maximum bearing area is presented to the soil to resist pull-out. Figure 2 illustrates the plate-like fluke and the "keying" action.

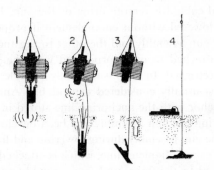

Touchdown   Penetration   Keying   Anchor established

Fig. 2.   Embedment and keying of a propellant-actuated anchor.

*For convenience the term "projectile" is sometimes used for "anchor-projectile".

In the second type of sediment projectile, two or more slender, movable flukes are hinged to the cylindrical body of the projectile. During penetration of the seafloor, they are clustered tightly about the body. Then, when a load is applied, they key by opening outward. Figure 3 shows this type of sediment projectile in the open position.

Existing projectiles for coral and rock do not have flukes. The projectile is shaped like a spear or arrowhead to achieve maximum penetration, and the lateral surfaces that engage the surrounding material can be serrated. Projectiles for use in coral and rock include a solid shaft with hardened point and serrated neck, a flat arrowhead shape, and a "three-dimensional" arrowhead (a pointed trihedron of flat plates with serrated or nonserrated edges). Figure 4 shows a coral rock type of anchor projectile.

The reaction vessel can be configured into practically any form that provides effective mass and high drag to minimize recoil and ensure optimum projectile velocity. It can be designed to entrap water to provide the mass (drag cones or plates are examples), or it can simply utilize the mass of the steel. The latter approach is less costly, but the resulting anchor system is heavier and recoil distances are greater. Reaction distances can vary from about 8 to 50 ft depending upon the reaction vessel configuration and the effective mass.

Several techniques are available for placing and firing propellant-actuated anchors. Such factors as the size and design of anchor, depth of water, handling equipment, and the overall operational requirements dictate the method to be used. The anchor can be fired by lowering it until a probe extending below it touches the bottom and triggers the firing mechanism. Or, the anchor can be held suspended above the bottom and fired by a signal from the surface through a firing line. The latter method is limited to depths less than about 200 ft and requires close control of movement of the surface work platform. A third method is to position the anchor on the seafloor by means of a support frame. In one design the reaction vessel also serves as a support frame. In this case, the anchor is properly oriented to fire the anchor perpendicular to the bottom without regard for bottom slope. Firing the anchor with a support frame is usually achieved by signal through a firing line from the surface. However, coded sound signals for firing the anchor are possible. Also, a touchdown sensor with a delay mechanism that permits the anchor to attain its proper orientation on the bottom before ignition has been used successfully.

Depending upon the mode of operation for a propellant-actuated anchor, up to three cables from the surface may be needed—the main anchor line, a line for retrieving the gun assembly, and an electrical cable for remote firing of the gun. In water less than 600 ft, two or more lines can be lowered without entanglement if proper precautions are taken. One line is attached to the gun assembly and the other to the anchor. The firing cable can be a separate line or be attached to or incorporated with the gun assembly line. After firing the gun assembly is retrieved. In deep water, only a single line can be lowered; as a result, the gun assembly is usually considered expendable. Another alternative in deep water is to free fall the anchor with the anchoring line stored in a bale on the anchor. A novel approach for retrieving the gun assembly has been developed by S. N. Marep.* A single line is attached to the gun assembly during lowering and firing. After firing, the gun assembly is retrieved, and a small diameter line, which is attached to the anchor downhaul cable (a short cable attached to the fluke) and gun assembly and located in the reaction case, is unreeled. The main anchoring line is placed over this guideline and lowered until it clamps to the downhaul cable. This technique is usable to depths as great as 3,000 ft

---

*The developer of the PACAN anchors, Sections 3.10 and 3.11.

2.1.2. *Advantages and disadvantages.* The principal advantages of propellant-actuated anchors are: (1) the anchor assembly is a compact package and has a higher holding capacity/weight efficiency than other anchors of the same capacity. (2) The anchor can function in a broad range of sediments and in material at least as hard as coral and vesicular basalt. (3) The concept is very nearly perfected. (4) Because penetration is rapid, special efforts to keep the surface vessel on station during embedment of the anchor are not required. (5) The light weight simplifies operational and handling difficulties.

The principal disadvantages are: (1) This type of anchor is not suited for a seafloor where there is rubble, medium-to-large-size boulders, pillow basalt, or rock overlain by sediment. (2) Special shipment, storage, and handling is required for the ordnance features. (3) The gun assembly is not generally retrievable in deep water. (4) The downhaul cable that subsequently becomes part of the anchor line is susceptible to abrasion and deterioration.

2.1.3. *History and status.* Propellant anchors were first developed in the late 1950s. Since then anchors ranging in nominal holding capacity from 1,000 to 220,000 lb have been developed and tested. Most of the anchors were designed for shallow-water applications (less than 600 ft), but some can be used in depths of more than 10,000 ft and a few were designed for operation to 20,000 ft.

Propellant-actuated anchors are still basically in their infancy. Considerably more testing and actual field use are required to develop user confidence in their unique capabilities and to eliminate the onus of fear and uncertainty that surrounds them.

## 2.2. *Vibrated direct-embedment anchors*

2.2.1. *Description.* A vibrated anchor (also referred to as "vibratory" anchor) is one that is driven into the seafloor by vibration. It is a long, slender metal construction consisting of a fluke-shaft assembly and a vibrator; for deep-water use (greater than 600 ft), a support guidance frame and a storage battery power pack are required. The deep-water system is illustrated in Fig. 5; the shallow-water system is shown in Section 3.1.3.

The vibrator that drives this type of anchor consists of counter-rotating eccentric masses* which can be either hydraulically driven from the surface or electrically powered at the seafloor.

The fluke used for both the Ocean Science and Engineering (OSE) anchor (shallow water) and the CEL anchor (deep water) is the special rotating Y-fluke developed under the CEL free-fall anchor program (Smith, 1966). A variety of sizes is available for the anchors. It has been shown both analytically and experimentally that a variety of sizes is necessary to utilize the available vibrator energy effectively. Also, anchor performance (penetration and resulting holding capacity) is dependent upon vibrator power, the supply of energy, the length of shaft, and seafloor properties.

The emplacement of this type of anchor consists of lowering the anchor assembly until it reaches the seafloor. The CEL bottom-resting system is activated upon bottom contact; the OSE system, which does not have a support frame, is activated prior to touchdown. The entire CEL anchor system is considered expendable in water depths greater than 1,000 ft. In lesser depths a second line can be used to retrieve the support frame. The OSE installation

---

*Linear accelerators have been designed, but greater success has been achieved with counter-rotating eccentrics.

technique allows retrieval of the vibrator unit after penetration is complete, because the anchor is lowered with dual lines with the anchor line being attached to the main shaft below the vibrator.

The CEL anchor has two additional features of interest—remote sensing instrumentation which permits determination of the attitude of the anchor when it rests on the seafloor, and a displacement monitoring system which yields penetration depth and rate.

2.2.2. *Advantages and disadvantages.* The principal advantages are: (1) It can accommodate layered seafloors or seafloors with variable resistances, because it has a continuous power application throughout penetration. (2) Penetration rate and amount can be monitored. (3) Confirmation of satisfactory implant is attainable. (4) Holding capacity is reasonably predictable.

The principal disadvantages are: (1) Use is limited to sediments. (2) It is difficult to handle from ship and stabilize on the seafloor. (3) The surface vessel must hold position precisely during penetration to prevent toppling. (4) Operation is limited to seafloors with slopes less than 10°.

2.2.3. *History and status.* The present designs function in sediments, attaining moderate holding capacities to water depths of 6,000 ft.

Pile and pipe driving by vibratory means has proven to be feasible on land and in water within the past several years. In 1967 the Ocean Science and Engineering Corporation successfully drove a coring pipe into the seafloor with a vibrator unit in 3,000 ft of water off Madagascar. This, combined with the CEL development of a quick-keying fluke (Smith, 1966) provided the catalyst for beginning work on the vibratory anchor concept. Since then, both surface- and seafloor-powered anchors have been designed and tested.

Recently, MKT Corporation and L. R. Foster Company have introduced hydraulic vibratory pile drivers usable to about 60 ft. However, with minor modifications, a depth of 1,000 ft should be attainable (Schmid, 1969). The feasibility of such a system has been demonstrated by the Institut Francais du Petrole (IFP) where a "Subsea Vibro-Driver" has been fabricated for use to depths of 650 ft (IFP, 1970). This device is designed to insert a large diameter core tube (12 in.) in sediments. It has been used occasionally to set stake piles for anchors.

### 2.3. *Screw-in anchors*

2.3.1. *Description.* A screw-in anchor (augured) is a slender shaft having one or more single-turn helical surfaces. It is, literally, screwed into the soil (see Section 3.14). This type of anchor was originally designed for use on land as a guy anchor for electrical transmission lines. New, suitable equipment has been developed to adapt it for use in the seabed. The primary application is as a pipeline anchor in shallow water. The diameter of the helixes, the number of helixes, the magnitude of downward force applied during penetration, the depth of penetration (by means of modular extensions to the shaft), the applied torque, and the strength of the shaft are varied to adjust to different soil properties.

2.3.2. *Advantages and disadvantages.* The principal advantages are: (1) Control of penetration. (2) Monitoring of penetration.

The principal disadvantages are: (1) Limited to use in shallow water. (2) Use is limited to sediments. (3) The surface vessel must hold precise position during installation.

2.3.3. *History and status.* Screw-in or augered anchors have only recently been introduced to the ocean environment; however, there was considerable land-based technology.

Adaptation for ocean use required only the development of a remote surface-powered driving unit. This anchor type is powered from the surface, and its water depth usage, therefore, depends upon properly transmitting power to the driving unit. The current usable water depth is limited to depths of several hundred feet. The principal uses are at present for pipelines in rather shallow depths (up to about 300 ft) in noncohesive soils.

## 2.4. Driven anchors

2.4.1. *Description.* A driven anchor is an anchor that is forced into the seabed by repeated impulsive loads, usually from a hammer. The particular forms are, at present, the stake pile (a single pile), the umbrella pile (a pile with fingerlike flukes that expand umbrella-fashion during the final phase of the driving), and a single-plate anchor that is driven with a mandrel and follower and then keyed by a pull-out load applied through the anchor line, as described for the propellant-actuated anchor. Figure 6 shows a stakepile and one type of umbrella pile. The top of a stake pile (the point of attachment of the anchor line) should be several feet below the seafloor, and the capacity of the stake pile to resist uplift is increased if the load on the pile has a horizontal component. Obviously, flukes minimize this requirement.

2.4.2. *Advantages and disadvantages.* The principal advantages are: (1) High capacity in sand. (2) Well established technique. (3) Maximum capacity attained with negligible movement (no keying or setting).

The principal disadvantages are: (1) Limited depth for surface air hammers (about 300 ft). (2) Limited depth for underwater hammers (about 1,000 ft). (3) Requires an enormous amount of surface support. (4) In the case of stake piles, the uplift-resisting capacity is reduced as the resultant load component approaches the vertical.

2.4.3. *History and status.* Present technology is limited to rather shallow depths (less than about 300 ft for surface-driven piles and 1,000 ft for underwater driving equipment) because of the present mechanical limitations of hammers and the large mass to be driven.

The state of the art of shallow-water driving is well advanced. Piles, fluked implements, and plates are commonly driven into the seafloor to provide uplift resistance. The driven plate is the most recent usage of the driving technique.

Driving from the surface is the most common and most advanced method of installing piles in the seafloor. Single-acting steam or compressed air hammers and diesel hammers are most often used. The present water depth record for driven piles is 340 ft; plans are underway to extend this record to 1,000 ft in the Santa Barbara Channel.

Subsurface driving is receiving considerable attention because the need for a long follower or expensive templates and surface support is reduced. Steam or compressed air hammers have been modified for underwater use, and have been utilized for pile driving in a water depth of 163 ft in Narragansett Bay, Rhode Island.

## 2.5. Drilled anchors

2.5.1. *Description.* A drilled anchor is a pile, a length of chain, or other structure that is placed into a hole previously drilled in the seafloor. (See Section 3.14 for an illustration). Methods for fixing the anchor in the hole include grout (and possibly a technique for expanding the grout against the sides of the hole) and mechanical ears or dogs that are forced outward to engage the sides of the hole when a pull-out load is applied. The technique is intended for rock and coral.

2.5.2. *Advantages and disadvantages*. The principal advantage is that it is virtually the only sure type of anchor for rock.

The principal disadvantage is that it requires close control of position during drilling.

2.5.3. *History and status*. Drilled and grouted anchors (piles and chains) provide reliable firm anchoring in seafloor rock and soil. Drilling is the only practicable method of emplacing piles in water depths in excess of 600 ft. Actually, drilling has been accomplished to a depth of 12,000 ft. The techniques are basically extensions of offshore oil-drilling methods.

Only a few vessels are available for emplacing pile anchors in very deep water. The Glomar Challenger, the drill ship for the Deep Sea Drilling Project, demonstrated a capability for installing pile anchors at a 20,000-ft depth. Other similar vessels could install piles to 6,000 ft. The major limitation of this anchoring technique is cost, which is up to $15,000/day.

Seafloor rock fasteners (such as rock bolts and grouted rebar), are presently limited to installation by diver and moderate holding capacities. Work to date has been involved with the techniques and equipment to install rock bolts and the shapes of the bolts for various materials. Descriptions and data are included in Section 3.19.

2.6. *Deadweight anchors*

2.6.1. *Description*. A deadweight anchor can be any object that is dense, heavy, and resistant to deterioration in water. It is the simplest and most crude form of an anchor. The type of ocean operation and the availability of materials usually dictate the shape, form, size, and weight of a deadweight anchor. Common examples of deadweight anchors are stones, concrete blocks, individual chain links, sections of chain links, and railroad wheels. (Figure 7 shows a primitive deadweight anchor.) Also, conventional drag-type anchors are sometimes used as deadweight anchors by themselves or in conjunction with other deadweight material.

In most instances a deadweight anchor functions as just that, a deadweight on the seafloor that resists uplift by its own weight in water and resists lateral displacement by its drag coefficient with the seafloor. Deadweight anchors are inefficient and unpredictable. Their drag coefficient varies with the amount of uplift force that coincides with lateral force. On sloping seafloors they tend to slide down slope or are displaced easily when the lateral force component is in the downslope direction. Deadweights are also easily displaced in shallow water by water drag from wave surge.

Conventional anchors are sometimes used as deadweight anchors to combat lateral movement. Of course, this application occurs only in water depths where it is impracticable to embed them by dragging. Conventional anchors used as deadweight anchors resist uplift force by their own weight and increase resistance to lateral displacement by as much as four times over a simple deadweight. A conventional anchor is used effectively in conjunction with simple deadweights by connecting it by chain or cable to the deadweight. The deadweight then provides the resistance to uplift, and the conventional anchor restricts the lateral displacement of the deadweight to a distance no greater than the chain length between them. In this application a much smaller conventional anchor can be used than if it were used alone.

2.6.2. *Advantages and disadvantages*. The principal advantages are: (1) They are simple to construct, economical, and readily available. (2) Their application is independent of most seafloor conditions, excluding steep, sloping bottoms. (3) Their uplift resistance is precisely predictable. (4) The installation procedures are relatively simple, and the installation equipment required is minimal.

The principal disadvantages are: (1) Their holding-capacity-to-weight ratio is undesirably low. (2) They become increasingly impracticable as holding capacity requirements extend beyond 1,500 lb. (3) They are highly susceptible to unpredictable lateral displacements. (4) They are costly to transport and handle because of their excessive weight.

## 3. DATA SUMMARIES

This section provides data on specific anchor designs that have been developed to meet special needs—primarily, the capability to resist uplift loads; the capability of being rapidly simply and precisely installed; and the capability of holding in hard material. Such requirements are not satisfied by anchors that must be preset by dragging.

The subsections summarize the data on specific anchor designs. Included are brief comments on the background or the area of use for the anchor, descriptions and details, operational aspects, and cost if established and known. While the details that are pertinent necessarily vary from anchor to anchor, those that are available are fitted into the following outline:

Source.
General Characteristics.
Details.
Operational Aspects.
Cost.
References.

Details on each anchor include such things as advertised nominal holding capacity, nominal penetration, operational depths, advantages and limitations. Care should be exercised in choosing an anchor based upon a company's advertised holding capacity because the capacity may not be necessarily based upon the same assumptions. Actual and estimated holding capacities are summarized in Appendix A and plotted in Appendix B.

Data on "operational modes" are pertinent in the case that more than one method of installation is available, because the method governs such things as speed of installation, precision of placement, and cost. For example, for propellant-actuated anchors, there are two ways to deliver the anchor to the seafloor (referred to as "free-fall" and "cable-lowered"), two ways to activate the firing mechanism ("automatic firing" upon contact with the seafloor and "command-firing" through manual operation of a switch aboard the surface vessel), and two options for dealing with ancillary equipment lowered with the anchor (to recover and reuse it or to abandon it). Of the eight combinations of these procedural options, more than one is often available.

Information not available at this writing, either because it was unknown or could not be obtained within the time frame for this writing, will be indicated by a dash mark. The date of preparation or latest revision is shown at the bottom of each page.

3.1. *Magnavox embedment anchor system, Model 1000 (propellant-actuated)*

3.1.1. *Source.* The Magnovox Company, 1700 Magnavox Way, Fort Wayne, Indiana 46804, U.S.A.

3.1.2. *General characteristics.* An operational, lightweight, compact, efficient, reliable anchor for use in automatically deployed moorings. It is: (1) suitable for free-fall, unguided, automatic placement, (2) adaptable to systems utilizing manual positioning and remote-command firing, (3) deployable in any depth, and (4) functional in a broad range of sea-floors.

*Advertised nominal holding capacity*

| | |
|---|---|
| Sandstone and coral: | 1,500 lb. |
| Sand | 2,000 lb. |
| Stiff clay | 1,200 lb. |
| Mud and soft clay | 500 lb. |

*Nominal penetration*

| | |
|---|---|
| Sand: | 10 ft. |
| Medium and stiff clays: | 6 to 12 ft. |
| Soft silt and clay: | 20 ft. |

*Water depth*

Design values—

| | |
|---|---|
| Maximum: | 20,000 ft. |
| Minimum: | 10 ft. |

Experience—

| | |
|---|---|
| Maximum: | 13,700 ft. |
| Minimum: | 10 ft. |

*Limitations*

No known limitations.

*Advantageous features*

Compact, functional unit.

Optional modes of operation, including operation with no line to the surface other than the anchor line.

3.1.3. *Details. Anchor assembly (excluding lines)*

With expendable gun assembly (see Fig. 8)—

| | |
|---|---|
| Height: | 3 ft. |
| Outside diameter: | 0.5 ft. |
| Weight: | 25 lb. |

With reusable gun assembly—

| | |
|---|---|
| Height: | 4 ft. |
| Outside diameter: | 2.0 ft. |
| Weight: | 100 lb. |

*Anchor projectile (see Figs. 9 and 10)*

Type:                Streamlined, compact projectile with elongated neck and pointed nose; four outward-opening flukes hinged to the projectile behind the shoulder.

FIG. 3. Anchor-projectile with hinged flukes extended.

FIG. 4. Three-finned anchor-projectile for coral seafloor.

Fig. 5. Deep-water vibrated anchor.

Fig. 6. Driven anchors.

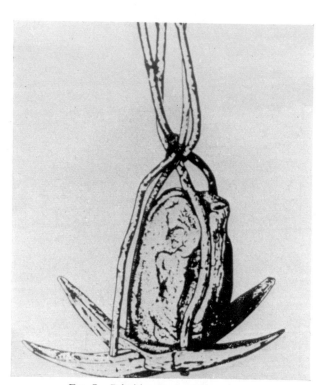

Fig. 7. Primitive deadweight anchor.

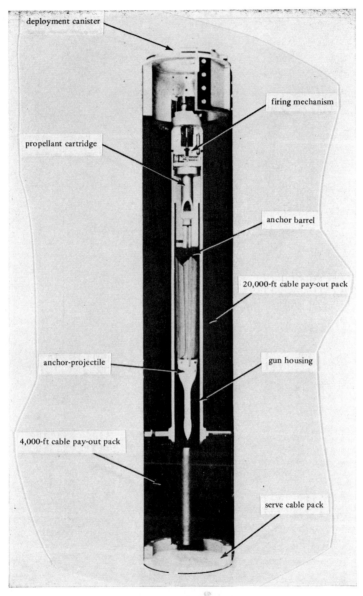

FIG. 8. Magnavox embedment anchor system, Model 1000; cutaway view of anchor-projectile and gun assembly mounted in expendable gun assembly.

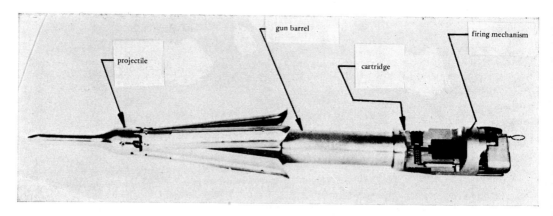

FIG. 9.  Magnavox embedment anchor system, Model 1000; without deployment canister.

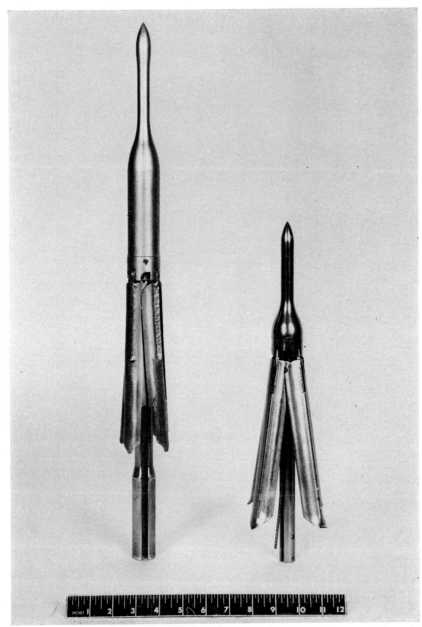

FIG. 10. Magnavox embedment anchor system; anchor-projectiles for Model 1000 (right) and Model 2000 (left).

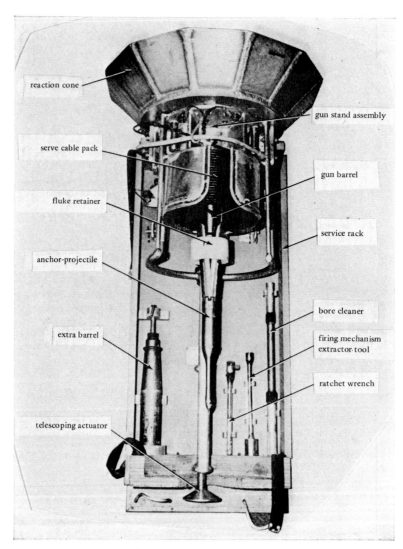

FIG. 11. Magnavox embedment anchor system, Model 2000; reusable gun assembly and accessories.

Length of projectile:                    16 in.
Diameter of projectile behind
    shoulder:                            1.50 in.
Length of fluke:                         8 in.
Width of fluke:                          1.25 in.
Effective area of flukes:                40 in².
Total weight:                            3.2 lb.

*Gun assembly* (see Fig. 9)
    Barrel diameter (inside):            0.75 in.
    Length of travel:                    8.5 in.
    Maximum working pressure:            —
    Separation velocity:                 —
    Upward reaction distance:            —
    Propellant:                          —
    Primer:                              —

*3.1.4. Operational aspects. Operational modes*
Expendable gun assembly (GA):
    1. Free-fall, automatic firing, GA not recovered.

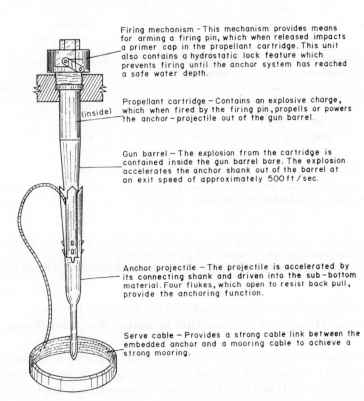

Firing mechanism – This mechanism provides means
for arming a firing pin, which when released impacts
a primer cap in the propellant cartridge. This unit
also contains a hydrostatic lock feature which
prevents firing until the anchor system has reached
a safe water depth.

Propellant cartridge – Contains an explosive charge,
which when fired by the firing pin, propells or powers
the anchor–projectile out of the gun barrel.

Gun barrel – The explosion from the cartridge is
contained inside the gun barrel bore. The explosion
accelerates the anchor shank out of the barrel at
an exit speed of approximately 500 ft / sec.

Anchor projectile – The projectile is accelerated by
its connecting shank and driven into the sub-bottom
material. Four flukes, which open to resist back pull,
provide the anchoring function.

Serve cable – Provides a strong cable link between the
embedded anchor and a mooring cable to achieve a
strong mooring.

FIG. 12.   Magnavox embedment anchor system, Model 2000; without reaction cone and gun
stand assembly.

Reusable gun assembly (GA):
1. Free-fall, automatic-firing, GA not recovered.
2. Free-fall, automatic-firing, GA recovered (primary mode).
3. Cable-lowered, automatic-firing GA not recovered.
4. Cable-lowered, automatic-firing, GA recovered.
5. Cable-lowered, command-firing, GA not recovered (unusual).
6. Cable-lowered, command-firing, GA recovered (unusual).

*Safety features*
Expendable gun assembly:
   Arming wire locks in-line/out-of-line piston in firing mechanism in the "safe"
      position—extracted just prior to launch.
   Spring-loaded in-line/out-of-line piston in firing mechanism—aligned when preset
      hydrostatic pressure is reached.
Reusable gun assembly:
   Arming wire—as above.
   Hydrostatic lock—as above.
   Hydrostatic lock on touchdown probe (Telescoping leg) prevents movement and
      triggering of firing mechanism.

3.1.5. *Cost.*

|                     | Material cost per anchor installation ($) | |
| Number of anchors   | Reusable gun assembly | Expendable gun assembly* |
| ------------------- | --------------------- | ------------------------ |
| 5                   | 460                   | 730                      |
| 10                  | 380                   | 720                      |
| 100                 | 280                   | 670                      |
| 500                 | 200                   | 520                      |
| 1,000               | 150                   | 370                      |

*Assumes one reusable gun assembly per 100 firings.

3.1.6. *References*
1. Excerpts from draft copy: Magnavox Self-Embedment Anchor Programs, 1962–1970. Fort Wayne, IN.
2. Letter, C. S. Myers (Magnavox) to R. J. Taylor (CEL), 18 Oct 1973.
3. The Magnavox Company. Brochure FWD539-1: The Magnavox Embedment Anchor System. Fort Wayne, IN., 1974.

3.2. *Magnavox embedment anchor system, Model* 2000* (*propellant-actuated*)
   3.2.1. *Source.* The Magnavox Company, 1700 Magnavox Way, Fort Wayne, Indiana 46804, U.S.A.
   3.2.2. *General characteristics.* An operational, lightweight, free-fall, propellant-actuated anchor for long-term (3 yr) mooring of small navigation buoys in sheltered water with currents of less than 3 knots. It can be: (1) deployed in water 10 ft deep, (2) installed by

*Data only available for Model 2000, which utilizes a reusable launching system.

one man, (3) carried on a 1/2-ton truck and on a small boat, and (4) embedded in a wide range of bottom material.

*Advertised nominal holding capacity*
Granite:                    1,500 lb.
Sandstone                   2,000 lb.
Coral:                      2,000 lb.
Sand:                       2,000 lb.
Stiff clay:                 1,700 lb.
Mud and soft clay:          800 lb.

*Nominal penetration*
Silty sand:                 10 to 12 ft.
Hard clay:                  10 to 12 ft.
Soft clay and silt:         18 to 20 ft.
Very soft silt:             26 to 30 ft.

*Water depth*
Design values—
  Maximum:                  —
  Minimum:                  10 ft.
Experience—
  Maximum:                  42 ft.
  Minimum:                  18 ft.

*Limitations*
No known limitations.

*Advantageous features*
Compact, functional unit.
Optional modes of operation, including operation with no line to the surface other than the anchor line.

3.2.3. *Details. Anchor assembly (excluding lines)* (see Fig. 11).
Height:                     4 ft.
Outside diameter:           2 ft.
Weight:                     110 lb.

*Anchor-projectile* (see Figs 12 and 13).
Type:                       Streamline, compact projectile with elongated neck and bulbous nose with ogibal point; four outward-opening flukes hinged to the projectile behind the shoulder.
Length of projectile:       25 in.
Diameter of projectile behind
  shoulder:                 1.5 in.

| Length of fluke: | 10 in. |
| Width of fluke: | 1.5 in. |
| Effective area of flukes: | 60 in². |
| Total weight: | 6.8 lb. |

*Gun assembly* (see Fig. 12).

| Barrel diameter (inside): | 1.13 in. |
| Length of travel: | 8.1 in. |
| Maximum working pressure: | 60,000 psi. |
| Separation velocity: | 500 ft/sec. |
| Upward reaction distance: | 2.5 ft. |
| Propellant: | 500 grains of Hercules HPC 87, 70 grains of Dupont IMR 3031, and 30 grains of Hercules No. 2400. |
| Primer: | Federal No. 215. |

3.2.4. *Operational aspects. Operational modes.*

With reusable gun assembly (GA):

1. Free-fall, automatic-firing, GA not recovered.
2. Free-fall, automatic-firing, GA recovered (primary mode).
3. Cable-lowered, automatic-firing, GA not recovered.
4. Cable-lowered, automatic-firing, GA recovered.
5. Cable-lowered, command-firing, GA not recovered.
6. Cable-lowered, command-firing, GA recovery (unusual).

*Safety features*

Arming wire locks in-line/out-of-line piston in firing mechanism in the "safe" position—extracted just prior to launch.

Spring-loaded in-line/out-of-line piston in firing mechanism—aligned when preset hydrostatic pressure is reached.

Hydrostatic lock on touchdown probe prevents telescoping and triggering until preset hydrostatic pressure is reached.

Shear pin in trigger lever shears if hydrostatic lock does not arm properly.

Visual indication of position of in-line/out-of-line piston.

3.2.5. *Cost.*

| Number of anchors | Material cost per anchor installation* ($) |
|:---:|:---:|
| 5 | 640 |
| 10 | 500 |
| 100 | 390 |
| 500 | 290 |
| 1,000 | 200 |

*Assumes one reusable gun assembly per 100 firings.

### 3.2.6. *References*

1. The Magnavox Company. Report No. FWD72-115: Explosive embedment anchor development program, by F. L. Erickson. Fort Wayne, IN., Nov 1972. (Contract No. DOT-GG-04468-A).
2. Letter, C. S. Myers (Magnavox) to R. J. Taylor (CEL), 18 Oct 1973.
3. The Magnavox Company. Brochure FWD539-1: The Magnavox Embedment Anchor System. Fort Wayne, IN., 1974.

## 3.3. *Vertohold embedment anchor*, 10K (*propellant-actuated*)

3.3.1. *Source*. Edo Western Corporation, 2645 South 2nd West, Salt Lake City, Utah 84115, U.S.A.

3.3.2. *General characteristics*. An operational, lightweight anchor for light-to-moderate duty (pipelines, tethered buoys, instruments, pontoon-bridge moorings) and for precise location of anchors in sand, stiff clay, or coral. It has been used for sewer outfalls and buoys.

*Advertised nominal holding capacity*
    10,000 lb.

*Nominal penetration*
    10 to 18 ft.

*Water depth*
    Design values—
        Maximum:                              —
        Minimum:                              —
    Experience—
        Maximum:                         1,100 ft.
        Minimum:                           45 ft.

*Limitations*
    —

*Advantageous features*
    Optional modes of operation, including operation with no line to the surface other than the anchor line.

3.3.3. *Details*. *Anchor assembly* (*excluding lines*) (see Figs 14 and 15)
    Height:                                2.5 ft.
    Maximum plan dimension
        (pendant container):               1.5 ft (estimated).
    Weight:                                60 lb.

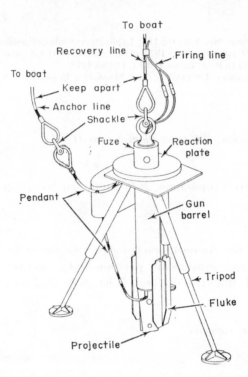

FIG. 16.   VERTOHOLD anchor assembly rigged for command-firing and recovery of gun
assembly.

*Anchor-projectile* (see Figs 13 and 14).

| | |
|---|---|
| Type: | Slightly tapered, solid shaft (projectile); two outward-opening flukes hinged to shaft at nose; flukes are flat plates with longitudinal stiffeners; two fluke sizes. |
| Length of projectile: | — |
| Length of fluke: | 14 in. |
| Width of fluke— | |
|     Anchor for soft material: | 5.5 in. |
|     Anchor for hard material: | 3.5 in. |
| Effective area of flukes— | |
|     Anchor for soft material: | 154 in$^2$. |
|     Anchor for hard material: | 98 in$^2$. |
| Total weight: | 25 lb. |

*Gun assembly*

| | |
|---|---|
| Barrel diameter (inside): | — |
| Length of travel: | — |
| Maximum working pressure: | — |
| Separation velocity: | — |
| Upward reaction distance: | — |

Propellant:                    0.34 lb of smokeless powder.
Primer:                        Shotgun shell.

### 3.3.4. *Operational aspects. Operational modes.*
1. Cable-lowered, command-firing, gun assembly recovered (see Fig. 16).
2. Cable-lowered, automatic-firing, gun assembly not recovered (see Fig. 17).

*Safety features*

Command-firing mode:

Safety pin in in-line/out-of-line detonator slide—removed before lowering assembly.

Hydrostatic-pressure actuation of in-line/out-of-line detonator slide.

Shorted-out electrical leads at surface.

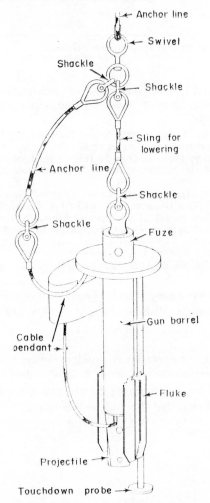

FIG. 17. **VERTOHOLD** anchor assembly rigged for automatic-firing and nonrecovery of gun assembly.

Automatic-firing mode:
    Safety pin in detonator slide.
    Hydrostatic-pressure actuation of detonator slide.
    Safety pin in touchdown probe to prevent movement—removed before lowering
    assembly.

### 3.3.5. *Cost.*

| Number of anchors | Material cost per anchor installation* ($) |
|---|---|
| 5 | 775 |
| 10 | 705 |
| 25 | 630 |
| 50 | 560 |
| 75 | 535 |
| 100 | 460 |

*Gun assembly is recovered; cost of the gun assembly not included.

### 3.3.6. *References*

1. Naval Civil Engineering Laboratory. Technical Report R-284-7: Structures in deep ocean; engineering manual for underwater construction, Chap. 7: Buoys and anchorage systems, by J. E. Smith. Port Hueneme, CA, Oct 1965. (AD473928).
2. Naval Civil Engineering Laboratory. Technical Note N-384: Investigation of embedment anchors for deep ocean use, by J. E. Smith. Port Hueneme, CA, Jul 1966.
3. Naval Civil Engineering Laboratory. Technical Note M-1133: Specialized anchors for the deep sea; progress summary, by J. E. Smith, R. M. Beard, and R. J. Taylor. Port Hueneme, CA, Nov 1970. (AD716408).
4. Naval Civil Engineering Laboratory. Technical Note N-1186: Explosive anchor for salvage operations; progress and status, by J. E. Smith. Port Hueneme, CA, Oct 1971. (AD735104).
5. Technical Note N-1186A: Addendum, by J. E. Smith. Port Hueneme, CA, Jan 1972.
6. Telephone conversation, Mr. Kidd (Edo Western) and Mr. Smith (CEL), 8 May 1972.
7. Edo Western Corporation. Report No. 13076: Operating procedures for Edo Western Corporation's Vertohold embedment anchor. Salt Lake City, UT, Sep 1972.
8. Edo Western Corporation. Pamphlet: Vertohold Embedment Anchors. Salt Lake City, UT, undated.

### 3.4. *Seastaple explosive embedment anchor, Mark* 5 (*propellant-actuated*)

3.4.1. *Source.* Teledyne Movible Offshore, Inc., P.O. Box 51936 O.C.S., Lafayette, Louisiana 70501, U.S.A.

3.4.2. *General characteristics.* An operational, rapidly emplaced, uplift-resisting anchor for precise placement in moderate depths and in any kind of seabed except very hard rock, and for various light-duty applications requiring direct embedment (no dragging for presetting the anchor), such as tiedowns and short-scope moorings.

*Advertised nominal holding capacity*
    5,000 lb.

*Nominal penetration*
    Coral:                        2 ft.
    Sand and medium clay:         7 ft.
    Mud and soft clay:            20 ft.

*Water depth*
    Design values—
        Maximum:                        1,000 ft.
        Minimum:                        10 ft.
    Experience—
        Maximum:                        6,000 ft.
        Minimum:                        10 ft.

*Limitations*
    Anchor not usable in rock seafloors.

*Advantageous features*
    Many expensive components are recoverable and reusable (optional).

3.4.3. *Details. Anchor assembly* (*excluding lines*) (see Fig. 18)
    Height without tripod and probe:     2.25 ft.
    Height with tripod or probe:         3 ft (estimated).
    Reaction cone diameter:              1.1 ft.
    Maximum plan dimension without
        tripod (pendant container):      1.5 ft (estimated).
    Diameter of tripod foot circle:      6 ft (estimated).
    Weight:                              60 lb.

*Anchor-projectile*
    Type:                                Rotating plate with keying flaps.
    Length overall:                      1.5 ft.
    Length of fluke:                     1.5 ft.
    Maximum width of fluke:              0.80 ft.
    Effective area of fluke:             0.83 ft$^2$.
    Total weight, including pendant:     10 lb.

*Gun assembly*
    Barrel diameter (inside):            —
    Length of travel:                    —
    Maximum working pressure:            10,000 psi.
    Separation velocity:                 —
    Upward reaction distance             —
    Propellant:                          0.125 lb.
    Primer:                              —

3.4.4. *Operational aspects. Operational modes.*
    1. Cable-lowered, automatic-firing, gun assembly not recovered (unusual).
    2. Cable-lowered, automatic-firing, gun assembly recovered (primary mode) (see
       Fig. 19).
    3. Cable-lowered, command-firing, gun assembly not recovered (unusual).
    4. Cable-lowered, command-firing, gun assembly recovered.

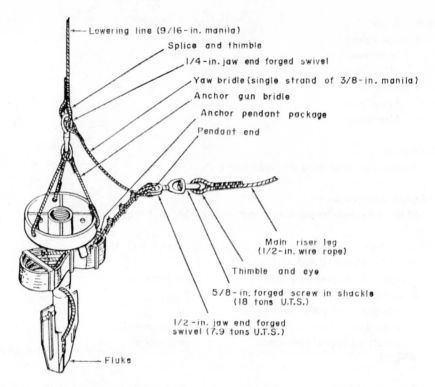

Lowering line (9/16-in. manila)
Splice and thimble
1/4-in. jaw end forged swivel
Yaw bridle (single strand of 3/8-in. manila)
Anchor gun bridle
Anchor pendant package
Pendant end

Main riser leg
(1/2-in. wire rope)

Thimble and eye

5/8-in. forged screw in shackle
(18 tons U.T.S.)

1/2-in. jaw end forged
swivel (7.9 tons U.T.S.)

Fluke

Fig. 19.   SEASTAPLE embedment anchor, Mark 5; rigged for recovery of gun assembly.

*Safety features*

Hydrostatic-pressure actuation of valve in firing mechanism.
Safety pin on hydrostatic pressure valve.
Shorting and grounding of electrical leads at upper end (command-firing mode).
Safety pin on sliding touchdown probe (automatic-firing mode).
Shielded electrical system.

3.4.5. *Cost.*

3.4.6. *References*

1. National Water Lift Company. Operation Instructions: seastaple anchor MK 5-4000. Kalamazoo, MI, Nov 1964.
2. Naval Ordnance Laboratory. Technical Report no. NOLTR 66-205: Field tests to determine the holding powers of explosive embedment anchors in sea bottoms, by J. A. Dohner. White Oak, MD, Oct 1966.
3. Naval Civil Engineering Laboratory. Technical Note N834: Investigation of embedment anchors for deep ocean use, by J. E. Smith. Port Hueneme, CA, Jul 1966.
4. J. L. Kennedy. "This lightweight explosion-set anchor can stand a big pull", Oil and Gas Journal, Vol 67, no. 16, Apr 21, 1969, pp. 84–86.
5. "New anchor penetrates rock bottoms", Offshore, Vol 29, no. 9, Aug 1969, pp. 104, 106–108.
6. Letter, C. D. Ellis (Movable Offshore, Inc.) to J. E. Smith (CEL), Sep 5, 1973.

3.5. *Seastaple explosive embedment anchor, Mark* 50 (*propellant-actuated*)

3.5.1. *Source.* Teledyne Movible Offshore, Inc., P.O. Box 51936 O.C.S., Lafayette, Louisiana 70501, U.S.A.

3.5.2. *General characteristics.* An operational, rapidly emplaced uplift-resisting anchor for precise placement in moderate depths and in any kind of seabed except very hard rock, and for short-scope anchoring in various offshore applications (vessels, rigs for offshore oil operations, etc.).

*Advertised nominal holding capacity*
    50,000 lb.

*Nominal penetration*
| | |
|---|---|
| Shale: | 4 ft. |
| Sand: | 20 ft. |
| Mud: | 40 ft. |

*Water depth*
    Design values—
| | |
|---|---|
|     Maximum: | 1,000 ft. |
|     Minimum: | 50 ft. |

    Experience—
| | |
|---|---|
|     Maximum: | 10,000 ft. |
|     Minimum: | — |

*Limitations*
    Not usable in competent rock seafloors.

*Advantageous features*
    Many expensive components are recoverable and reusable (optional).

3.5.3. *Details: Anchor assembly* (*excluding lines*) (see Fig. 20).
| | |
|---|---|
| Height without tripod or probe: | 8 ft. |
| Height with tripod or probe: | 10 ft (estimated). |
| Reaction cone diameter: | 4 ft. |
| Maximum plan dimension without tripod (pendant container): | 4 ft (estimated). |
| Diameter of tripod foot circle: | 18 ft (estimated). |
| Weight: | 1,900 lb. |

*Anchor-projectile*
| | |
|---|---|
| Type: | Rotating plate with keying flaps. |
| Length overall: | 7.5 ft (approx). |
| Length of fluke: | 4.5 ft (approx). |
| Maximum width of fluke: | 2.0 ft (approx). |

Effective area of fluke:            8.3 ft².

Total weight, including pendant:     250 lb.

*Gun assembly*

Barrel diameter (inside):         5 in.

Length of travel:                38 in.

Maximum working pressure:      —

Separation velocity:            450 ft/sec.

Upward reaction distance:       —

Propellant:                    3.5 lb.

Primer:                       —

### 3.5.4. *Operational aspects. Operational modes.*

1. Cable-lowered, automatic-firing, gun assembly not recovered (unusual).
2. Cable-lowered, automatic-firing, gun assembly recovered (primary mode).
3. Cable-lowered, command-firing, gun assembly not recovered (unusual).
4. Cable-lowered, command-firing, gun assembly recovered.

*Safety features*

Hydrostatic-pressure actuation of switch valve in firing mechanism.

Safety pin on sliding touchdown probe (automatic-firing mode).

Shielded electrical system.

### 3.5.5. *Cost.*

—

### 3.5.6. *References*

1. Army Mobility Equipment Research and Development Center. Report no. 1909-A: Development of multi-leg mooring system, Phase A. Explosive embedment anchor, by J. A. Christians and E. D. Meisburger. Fort Belvoir, VA, Dec 1967.
2. J. L. Kennedy. "This lightweight explosion-set anchor can stand a big pull", Oil and Gas Journal, Vol 67, no. 16, Apr 21, 1969, pp. 84–86.
3. "New anchor penetrates rock bottoms", Offshore, Vol 29, no. 9, Aug 1969, pp. 104, 106–108.
4. Letter, C. D. Ellis (Movible Offshore, Inc.) to J. E. Smith (CEL), Sep 5, 1973.

## 3.6. *Cel 20K propellant anchor (propellant-actuated)*

3.6.1. *Source.* Civil Engineering Laboratory, Naval Construction Battalion Center, Port Hueneme, California 93043, U.S.A.

3.6.2. *General characteristics.* An operational, direct-embedment anchor system of minimum cost for use in very deep water in short-scope moorings and other applications requiring significant resistance to uplift (see Fig. 21).

*Advertised nominal holding capacity*

20,000 lb.

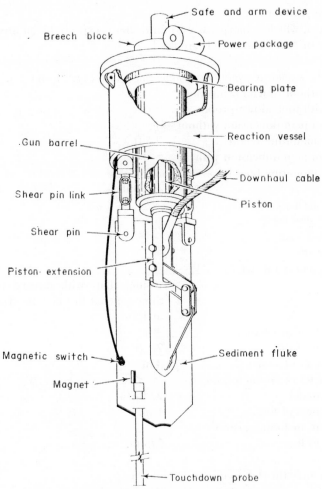

FIG. 22. Schematic of CEL 20K Propellant Anchor.

*Nominal penetration*
| | |
|---|---|
| Basalt: | 2 ft. |
| Sand: | 20 ft. |
| Medium clay: | 40 ft. |

*Water depth*
Design values—
| | |
|---|---|
| Maximum: | 20,000 ft. |
| Minimum: | 90 ft. |

Experience—
| | |
|---|---|
| Maximum: | 18,600 ft. |
| Minimum: | 50 ft. |

*Limitations*
No known limitations.

*Advantageous features*

The system, which is inexpensive to fabricate, is expendable in deep water. Surplus Army or Navy gun barrels are used.

3.6.3. *Details. Anchor assembly* (*excluding lines*) (see Figs 21 and 22).

| | |
|---|---|
| Height without touchdown probe: | 7 ft*. |
| Height with touchdown probe: | 9 ft*. |
| Maximum plan dimension without cable-mounting board: | 2.5 ft. |
| Maximum plan dimension with cable-mounting board | 3.5 ft. |
| Weight: | 1,800 lb.† |

*Add 2 ft for mud fluke.
†Add 200 lb for mud fluke.

*Anchor-projectile*

For rock and coral (see Fig. 23)—

| | |
|---|---|
| Type: | Round shaft with tapered nose (1 : 6) and three tapered fins (1 : 2)- Primary fins spaced at 140°. |
| Length of projectile: | 3 ft. |
| Length of fins: | 2.5 ft. |
| Diameter of shaft: | 3 in. |
| Diameter of circumscribing cylinder: | 27 in. |
| Thickness of fins: | 1 in. |
| Weight, including piston (115 lb): | 275 lb. |

For sand and stiff clay (see Figs 24 and 25)—

| | |
|---|---|
| Type: | Rotating plate. |
| Length overall: | 38 in. |
| Length of fluke: | 38 in. |
| Width of fluke: | 18 in. |
| Effective area of fluke: | 4.5 ft$^2$. |
| Total weight, including piston: | 300 lb. |

For sand and clay (2 × 4 ft)—

| | |
|---|---|
| Type: | Rotating plate. |
| Length overall: | 51 in. |
| Length of fluke: | 51 in. |
| Width of fluke: | 24 in. |
| Effective area of fluke: | 8.0 sq ft. |
| Total weight, including piston: | 370 lb. |

For mud and soft clay (2-1/2 × 5 ft)—

| | |
|---|---|
| Type: | Rotating plate. |
| Length overall: | 63 in. |

Length of fluke:                           63 in.
Width of fluke:                            30 in.
Effective area of fluke:                   12.5 ft$^2$.
Total weight, including piston:            490 lb.

*Gun assembly*
Barrel diameter (inside):                  4.25 in.
Length of travel:                          26 in.
Maximum working pressure:                  35,000 psi.
Separation velocity:                       400 ft/sec.
Upward reaction distance:                  25 ft.
Propellant:                                3.75 lb max of Standard Navy pyrotechnic
                                           (smokeless).
Primer:                                    M-58 (black powder).

*3.6.4. Operational aspects. Operational modes.*
1. Cable-lowered, automatic-firing, gun assembly not recovered (see Fig. 21).
2. Cable-lowered, automatic-firing, gun assembly recovered.

*Safety features*
Safety pin holds in-line/out-of-line plunger in safe-and-arm device out of line—
    extracted prior to lowering.
Hydrostatic-pressure actuation of in-line/out-of-line plunger.
Hydrostatic-pressure actuation of switch in power package.

*3.6.5. Cost.* The material cost per anchor installation when purchased in lots of from
1 to 20 anchors is:
$1,360                                     With reusable gun assembly (assumes one re-
                                           usable gun assembly per 20 firings; the gun
                                           assembly cost is $3,200).
$4,500                                     With expendable gun assembly.

*3.6.6. References*
1. Naval Civil Engineering Laboratory. Technical Note N-1282: Propellant-actuated deep water anchor;
   interim report, by R. J. Taylor and R. M. Beard. Port Hueneme, CA, Aug 1973. (AD765570).

*3.7. Cel 100K propellant anchor (propellant-actuated)*

*3.7.1. Source.* Civil Engineering Laboratory, Naval Construction Battalion Center, Port
Hueneme, California 93043, U.S.A.

*3.7.2. General characteritsics.* An operational anchor that is undergoing further testing
of a revised launch vehicle design and a new sediment anchor-projectile design. The anchor,
which is for use in ship-salvage operations, (1) can be placed rapidly without dragging
(2) develops the full working strength of a standard Navy beach gear leg, and (3) can be
handled on an ARS, ASR, ATF, or ATS.

*Advertised nominal holding capacity*
    100,000 lb.

*Nominal penetration*
    Vesicular basalt:                  2 ft.
    Coral:                               7 ft.
    Sand:                              20 to 30 ft.
    Mud:                              30 to 50 ft.

*Water depth*
    Design values—
        Maximum:                500 ft.
        Minimum:                50 ft.
    Experience—
        Maximum:                700 ft.
        Minimum:                35 ft.

*Limitations*
    Seafloor must be level and smooth enough to assure upright attitude of launch vehicle (tilt less than 30°).
    Potential entanglement problems with multiple lines.

*Advantageous features*
    Stable, rugged launch vehicle.
    Total manual control of placement and firing, which permits interruptions to assure correctness of operation.
    High capacity in sand, coral, and rock.
    Many expensive components are recoverable and reusable.

3.7.3. *Details. Anchor assembly* (*excluding lines*) (see Fig. 26).
    Height:                          11 ft.
    Plan dimension:              8 ft$^2$.
    Weight:                          13,000 lb.

*Anchor-projectile*
    For rock and coral—
        Type:                     Three fixed fins (Y-section with 120-degree dihedral angles); fins tapered.
        Length of projectile:       6.75 ft.
        Length of fins:           5.0 ft.
        Diameter of circumscribing
            cylinder:                3.1 ft.
        Weight, including piston
            (500 lb):               2,000 lb.
    For sand and coral (new design) (see Fig. 27)—
        Type:                     Rotating plate.

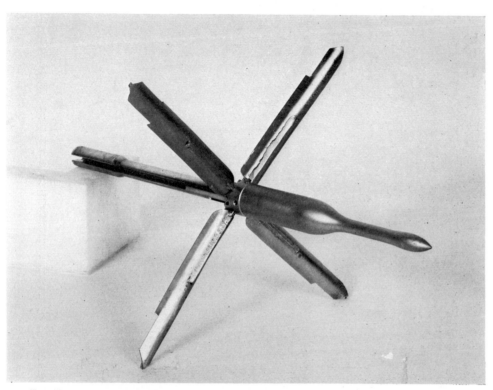

Fig. 13. Magnavox embedment anchor system, Model 2000; anchor with flukes fully opened.

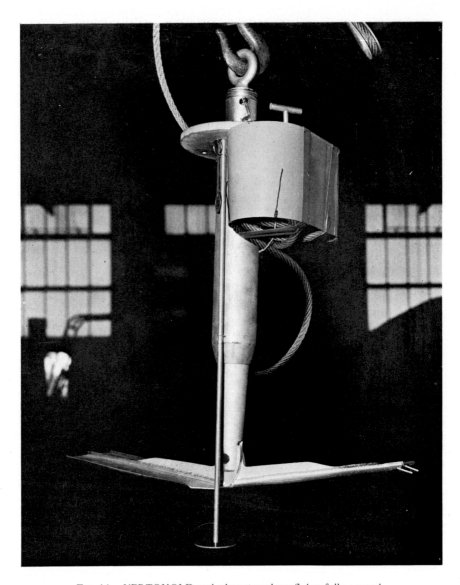

FIG. 14.   VERTOHOLD embedment anchor; flukes fully opened.

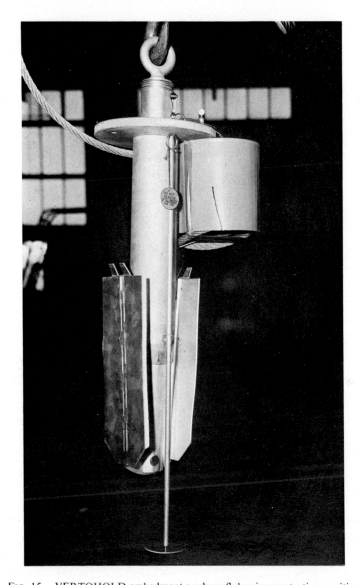

Fig. 15.   VERTOHOLD embedment anchor; flukes in penetrating position.

drag cone

gun

anchor fluke

touchdown probe

Fig. 18. SEASTAPLE embedment anchor, Mark 5.

FIG. 20. SEASTAPLE embedment anchor, Mark 50.

FIG. 21.   CEL 20K Propellant Anchor.

FIG. 23. CEL 20L Propellant Anchor; rock fluke and piston.

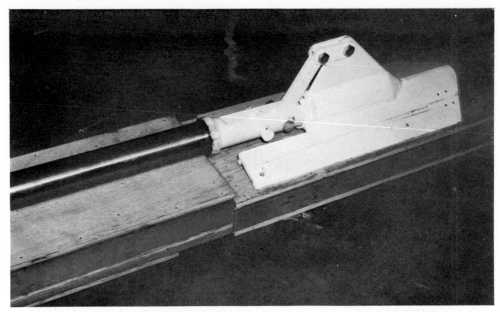

FIG. 24. CEL 20K Propellant Anchor; sand fluke and piston in penetrating position.

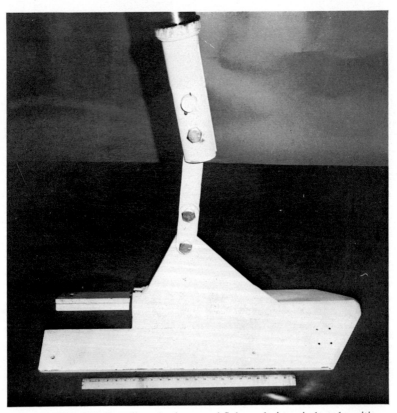

FIG. 25. CEL 20K Propellant Anchor; sand fluke and piston in keyed position.

FIG. 26.   CEL 100K Propellant Anchor; launch vehicle with dummy anchor (used to evaluate gun performance).

FIG. 27.   CEL 100K Propellant Anchor; flukes.

FIG. 28.    Army explosive embedment anchor, XM-50; front quarter view.

FIG. 29.    Army explosive embedment anchor, XM-50; rear quarter view.

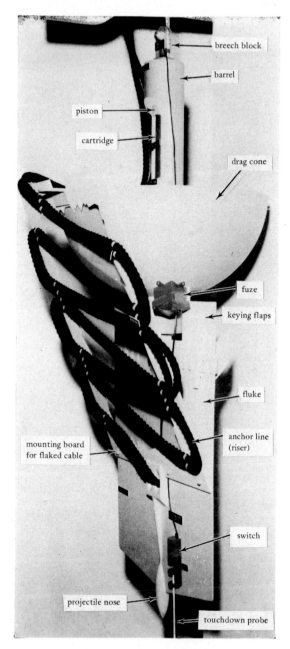

breech block

barrel

piston

cartridge

drag cone

fuze

keying flaps

fluke

mounting board
for flaked cable

anchor line
(riser)

switch

projectile nose

touchdown probe

FIG. 30. Army explosive embedment anchor, XM-200; cutaway model.

FIG. 31. Army explosive embedment anchor, XM-200; front quarter view.

FIG. 32. Army explosive embedment anchor, XM-200; rear quarter view.

FIG. 33.   PACAN 3DT; equipped with plate-type fluke mounted in cradle aboard ship.

FIG. 34.   PACAN 3DT; anchor-projectiles for sediments.

Fig. 36. PACAN 10DT; without anchor projectile.

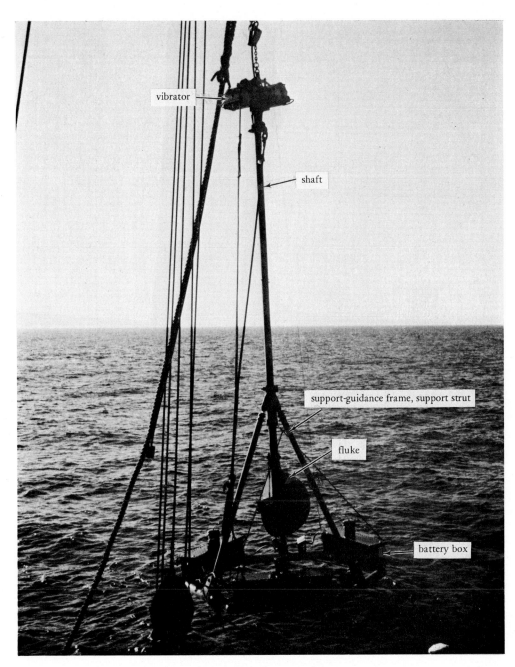

vibrator

shaft

support-guidance frame, support strut

fluke

battery box

FIG. 37.   Navy vibratory anchor.

FIG. 38.   Navy vibratory anchor; quick-keying fluke shown in position assumed after keying.

| Length of projectile: | 5-1/2 ft. |
| Length of fluke: | 5/1/2 ft. |
| Width of fluke: | 2-3/4 ft. |
| Effective area of fluke: | 13 ft². |
| Weight, including piston (500 lb): | 1,550 lb. |

For clay (new design) (see Fig. 27)—

| Type: | Rotating plate. |
| Length of projectile: | 6-2/3 ft. |
| Length of fluke: | 6-2/3 ft. |
| Width of fluke: | 3-1/3 ft. |
| Effective area of fluke: | 22 sq ft. |
| Weight, including piston (500 lb): | 1,900 lb. |

*Gun assembly*

| Barrel diameter (inside) | 10 in. |
| Length of travel: | 36 in. |
| Maximum working pressure: | 35,000 psi. |
| Separation velocity: | 350 to 400 ft/sec. |
| Upward reaction distance: | 8 to 12 ft. |
| Propellant: | 14 lb of M26 smokeless powder. |
| Primer: | M58 (black powder). |

### 3.7.4. *Operational aspects. Operational modes.*
1. Cable-lowered, command-firing, launch vehicle recovered.

*Safety features*

Lanyard-operated safety pin—pulled as launch vehicle leaves the deck.

Interrupted explosive train, with in-line/out-of-line plunger controlled by switch aboard ship and by hydrostatic pressure.

Visual safe–arm indication.

### 3.7.5. *Cost.*
The material cost per anchor installation when purchased in lots of from 1 to 20 anchors is $4,100. This assumes one reusable gun assembly per 20 firings with the gun assembly cost being $17,000. The piston, which costs $1,500, is recovered 50% of the time. This cost also includes $1,000 for an expendable anchor pendant.

### 3.7.6. *References*

1. Naval Ship Systems Command. Supervisor of Salvage. NAVSHIPS 0994-007-1010: Technical manual: Assembly, stowage, and operation; Anchor, salvage embedment. Washington, DC, Jan 1970.
2. Naval Civil Engineering Laboratory. Technical Note N-1186: Explosive anchor for salvage operations; progress and status, by J. E. Smith. Port Hueneme, CA, Oct 1971. (AD735104).
3. Technical Note N-1186A: Addendum, by J. E. Smith. Port Hueneme, CA, Jan 1972.

### 3.8. *Explosive embedment anchor, XM-50 (propellant-actuated)*

3.8.1. *Source.* U.S. Army Mobility Equipment Research and Development Center, Code SMEFB-HP, Fort Belvoir, Virginia 22060, U.S.A.

3.8.2. *General characteristics.* An operational, lightweight anchor (relative to a conventional drag-type anchor of comparable capacity) that uses rope instead of chain in multi-leg moorings for 25,000-DWT tankers in shallow, exposed coastal waters (maximum wave height, 11 ft). It is (1) reliable, (2) quickly installed by Army personnel, (3) suitable for any kind of seafloor material except consolidated rock, (4) adaptable to fleet-type single-point moorings, (5) and air-transportable (C-130).

*Advertised nominal holding capacity*
    50,000 lb.

*Nominal penetration*
| | |
|---|---|
| Coral: | 20 ft. |
| Sand: | 20 ft. |
| Mud and soft clay: | 40 ft. |

*Water depth*
| | |
|---|---|
| Design values— | |
|     Maximum: | 150 ft. |
|     Minimum: | 25 ft. |
| Experience— | |
|     Maximum: | 51 ft. |
|     Minimum: | 9 ft. |

*Limitations*
    Not usable in competent rock seafloors.

*Advantageous features*
    Many expensive components are recoverable and reusable.

3.8.3. *Details. Anchor assembly (excluding lines) (see Figs 28 and 29).*
| | |
|---|---|
| Height, including probe extension (2.5 ft): | 12.2 ft. |
| Drag cone diameter: | 4.0 ft. |
| Weight, including riser cable (70 lb): | 1,900 lb. |

*Anchor-projectile (see Figs 28 and 29).*
| | |
|---|---|
| Type: | Rotating plate. |
| Length overall: | 4.83 ft. |
| Length of fluke: | 4.83 ft. |
| Width of fluke: | 2.0 ft. |
| Effective area of fluke: | 8 ft$^2$. |
| Weight (includes piston): | 400 lb. |

*Gun assembly*

| | |
|---|---|
| Barrel diameter (inside): | 5 in. |
| Length of travel: | 38 in. |
| Maximum working pressure: | 53,000 psi. |
| Separation velocity: | 400 to 500 ft/sec. |
| Upward reaction distance: | 10 ft. |
| Propellant: | 3.5 lb of MS (MIL-P-323). |
| Primer: | Two WOX69A (Navy Mk 101) and 6 to 7 ft of Dupont Pyrocore no. 2040 cord. |

**3.8.4.** *Operational aspects. Operational modes*
Cable-lowered, automatic-firing, gun assembly recovered.

*Safety features*
Hydrostatic-pressure actuation of switch in fuse.

**3.8.5.** *Cost.* The material cots per anchor installation when purchased in lots of from 1 to 20 anchors is $4,750. This assumes one reusable gun assembly per 20 firings with the gun assembly costing $3,000. The polyurethane-coated Nylon pendant, costing $1,000, is expendable.

**3.8.6.** *References*
1. Army Mobility Equipment Research and Development Center. Report No. 1909-A: Development of multi-leg mooring system, Phase A. Explosive embedment anchor, by J. A. Christians and E. P. Meisburger. Fort Belvoir, VA, Dec 1967.
2. Letter, Commander U.S. Army CDC to Distribution H, 1 Nov 1972, subject: Revised Department of the Army Approved Qualitative Material Requirement (QMR) for multi-leg tanker mooring system.
3. H. C. Mayo. "Explosive anchors for ship mooring", Marine Technology Society Journal, Vol 7, no. 6, Sep 1973, pp 27–34.
4. Army Mobility Equipment Research and Development Center. Report No. 2078: Explosive embedment anchors for ship mooring, by H. C. Mayo. Fort Belvoir, VA, Nov 1973.
5. Letter, Commander U.S. Army MERDC to Commander NCEL, 25 Feb 1974, subject: MERDC explosive anchor.

**3.9.** *Explosive embedment anchor, XM-200 (propellant-actuated)*

**3.9.1.** *Source.* U.S. Army Mobility Equipment Research and Development Center, Code SMEFB-HP, Fort Belvoir, Virginia 22060, U.S.A.

**3.9.2.** *General characteristics.* An operational anchor that uses rope instead of chain in multileg moorings for 40,000-DWT tankers in shallow, sheltered coastal waters (maximum wave height, 3 ft). The anchor is (1) reliable, (2) quickly installed by Army personnel, (3) suitable for any kind of seafloor material except consolidated rock, (4) adaptable for fleet-type single-point moorings, and (5) a component of a mooring system that is a significantly smaller logistic burden than systems using conventional drag-type anchors.

*Advertised nominal holding capacity*
200,000 lb.

*Nominal penetration*
    Coral:                                          15 ft.
    Sand and stiff clay:               20 ft.
    Mud and soft clay:            30 to 40 ft.

*Water depth*
    Design values—
        Maximum:                   150 ft.
        Minimum:                   40 ft.
    Experience—
        Maximum:                   55 ft.
        Minimum:                   36 ft.

*Limitations*
    Not usable in competent rock seafloors.

*Advantageous features*
    Many expensive components are recoverable and reusable.

3.9.3. *Details. Anchor assembly* (*excluding lines*) (see Fig. 30).
    Height, including probe extension
        (4.0 ft):                     18.0 ft.
    Drag cone diameter:          4.0 ft.
    Weight, including riser cable
        (1,200 lb):                5,300 lb.

*Anchor-projectile* (see Figs 30 and 31).
    Type:                                Rotating plate.
    Length overall:                6.6 ft.
    Length of fluke:             6.6 ft.
    Width of fluke:               3.5 ft.
    Effective area of fluke:      20 ft$^2$.
    Weight (including piston):   1,200 lb.

*Gun assembly* (see Fig. 32).
    Barrel diameter (inside):     6 in.
    Length of travel:            60 in.
    Maximum working pressure:  60,000 psi.
    Separation velocity:        400 ft/sec.
    Upward reaction distance:   30 ft.
    Propellant:                   14 lb of Pyrocellulose (Navy 8/55 smokeless).
    Primer:                       Two WOX69A (Navy Mk 101) and 9 ft of
                                      Dupont Pyrocore No. 2040 cord.

4.9.3. *Operational aspects. Operational modes.*
    Cable-lowered, automatic-firing, gun assembly recovered.

*Safety features*
Hydrostatic-pressure actuation of switch in fuse.

3.9.5. *Cost.* The material cost per anchor installation when purchased in lots of from 1 to 20 anchors is $11,450. This assumes one reusable gun assembly per 20 firings with the gun assembly costing $9,000. The polyurethane-coated Nylon pendant, costing $1,000, is expendable.

3.9.6. *References*

1. Army Mobility Equipment Research and Development Center. Report No. 1909-A: Development of multi-leg mooring system, Phase A. Explosive embedment anchor, by J. A. Christians and E. P. Meisburger. Fort Belvoir, VA, Dec 1967.
2. Letter, Commander U.S. Army CDC to Distribution H, 1 Nov 1972, subject: Revised Department of the Army approved Qualitative Material Requirement (QMR) for multi-leg tanker mooring system.
3. H. C. Mayo. "Explosive anchors for ship mooring", Marine Technology Society Journal, Vol 7, no. 6, Sep 1973, pp 27–34.
4. Army Mobility Equipment Research and Development Center. Report No. 2078: Explosive embedment anchors for ship mooring, by H. C. Mayo. Fort Belvoir, VA, Nov 1973.
5. Letter, Commander U.S. Army MERDC to Commander NCEL, 25 Feb 1974, subject: MERDC explosive anchor.

3.10. *Pacan 3DT (propellant-actuated).*

3.10.1. *Source.* Union Industrielle et d'Enterprise, 49 bis, Avenue Hache, 75005 Paris, France.

3.10.2. *General characteristics.* An operational anchor whose known testing has been confined to corals and shelly limestone (30 installations). It was designed as a mooring anchor for both sediments and rock (see Fig. 33).

*Advertised nominal holding capacity*
  66,000 lb.

*Nominal penetration*
  —

*Water depth*
  Design values—
    Maximum:                    3,000 ft and 20,000 ft (two designs).
    Minimum:                    —
  Experience—
    Maximum:                    300 ft.
    Minimum:                    —

*Limitations*
  —

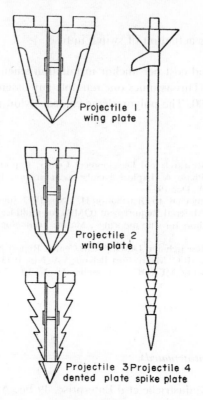

Projectile I
wing plate

Projectile 2
wing plate

Projectile 3 Projectile 4
dented plate spike plate

Fig. 35.   PACAN 3DT; anchor-projectiles.

*Advantageous features*

Many expensive components are recoverable and reusable.

A special auxiliary connector in the anchor line (optional) is designed to permit recovery of the gun assembly without line entanglement, facilitate emplacement of heavy mooring lines, and permit installation of the mooring some time after installation of the anchor.

3.10.3.   *Details. Anchor assembly* (see Fig. 33).

| | |
|---|---|
| Height with spike projectile, including probe: | 25 ft. |
| Height with plate projectile, including probe: | 16 ft (estimated). |
| Outside diameter: | 4.6 ft. |
| Weight, including plate projectile and drum for pendant: | 5,300 lb. |

*Anchor-projectiles*

For sand (see Figs 34 and 35)—

Type:                          Rotating plate of arrowhead shape, with stiffening ribs.

| | |
|---|---|
| Length of projectile: | 4.7 ft. |
| Length of fluke: | 4.7 ft. |
| Width of fluke: | 2.4 ft. |
| Effective area of fluke: | 7.4 ft$^2$. |
| Weight: | 800 lb. |
| For rock (see Fig. 35)— | |
| Type: | Spike. |
| Length: | 17 ft (estimated). |
| Weight: | — |

*Gun assembly*

| | |
|---|---|
| Barrel diameter (inside): | 4 in (approx) |
| Length of travel: | — |
| Maximum working pressure: | — |
| Separation velocity: | — |
| Upward reaction distance: | — |
| Propellant: | — |
| Primer: | — |

3.10.4. *Operational aspects. Operational modes.*
   1. Cable-lowered, automatic-firing, gun assembly recovered (primary mode).
   2. Cable-lowered, command-firing, gun assembly recovered.

*Safety features*
   Hydrostatic-pressure arming.
   Shorting of leads at surface vessel (optional, command-firing mode).

3.10.5. *Cost.* The material cost per anchor installation when purchased in lots of from 1 to 20 anchors is $4,570. This assumes one reusable gun assembly per 20 firings with the gun assembly costing $17,300. Note: Cost will vary $\pm$$500 per anchor according to type of fluke. Cost figures include 2-1/2% charge for packaging for export. Cost figures pertain to gun designed for 3,000-ft depth; add approximately $5,700 for gun designed for 20,000 ft.

3.10.6. *References*
1. Letter, P.D.L. (MAREP) to W. J. Tudor (NAVFAC), Jul 22, 1969.
2. Letter, J. Liautaud (UiE) to R. J. Taylor (CEL), Sep 11, 1973.

3.11. *Pacan 10DT (propellant-actuated).*

3.11.1. *Source.* Union Industrielle et d'Enterprise, 49 bis, Avenue Hache, 75008 Paris, France.

3.11.2. *General characteristics.* This anchor has been fabricated, but is untested. It was designed as a large-capacity anchor for sediments and rock (see Fig. 36).

*Advertised nominal holding capacity*
   220,000 lb.

*Nominal penetration*
—

*Water depth*
 Design values—
  Maximum:      3,000 ft.
  Minimum:      —
 Experience—
  Maximum:      Not tested.
  Minimum:      Not tested.

*Limitations*
—

*Advantageous features*
 Many expensive components are recoverable and reusable.
 A special auxiliary connector in the anchor line (optional) is designed to permit recovery of the gun assembly without line entanglement, facilitate emplacement of heavy mooring lines, and permit installation of the mooring some time after installation of the anchor.

3.11.3. *Details. Anchor assembly* (see Fig. 36).
 Height with spike projectile,
  including probe:     44 ft.
 Height with plate projectile,
  including probe:     31 ft (estimated).
 Outside diameter:     7.2 ft.
 Weight, including plate projectile
  and drum for pendant:   19,400 lb.

*Anchor-projectile*
 For sand (see Figs 34 and 35)—
  Type:       Rotating plate of arrowhead shape, with stiffening ribs.
  Length of projectile:   —
  Length of fluke:    9.3 ft.
  Width of fluke:    3.0 ft.
  Effective area of fluke:  18 ft$^2$.
  Weight:      3,000 lb (estimated).
 For rock (see Fig. 35)—
  Type:       Spike.
  Length:      —
  Weight:      —

*Gun assembly*
 Barrel diameter (inside):  8 in. (approx).

Length of travel:                     —
Maximum working pressure:             —
Separation velocity:                  —
Upward reaction distance:             —
Propellant:                           —
Primer:                               —

3.11.4. *Operational aspects. Operational modes.*
1. Cable-lowered, automatic-firing, gun assembly recovered (primary mode).
2. Cable-lowered, command-firing, gun assembly recovered.

*Safety features*
—

3.11.5. *Cost.* The material cost per anchor installation when purchased in lots of from 1 to 20 anchors is $12,570. This assumes one reusable gun assembly per 20 firings with the gun assembly costing $29,400. Note: Costs are approximate values for plate projectiles. Add approximately $1,300 for spike anchors. Cost figures include 2-1/2 % charge for packaging for export.

3.11.6 *References*
1. Letter, P.D.L. (MAREP) to W. J. Tudor (NAVFAC) Jul 22, 1969.
2. Letter, J. Liautaud (UiE) to R. J. Taylor (CEL), Sep 11, 1973.

3.12. *Direct-embedment vibratory anchor (vibrated).*

3.12.1. *Source.* Civil Engineering Laboratory, Naval Construction Battalion Center, Port Hueneme, California 93043, U.S.A.

3.12.2. *General characteristics.* An operational, reliable anchor for direct embedment in all sediments and in water depths to 6,000 ft. It combines low cost (components either inexpensive or recoverable) with lightweight, and it develops holding capacities for loads in any direction.

*Advertised nominal holding capacity*
    Sand:                             40,000 lb.
    Clay:                             25,000 lb.

*Nominal penetration*
    Sand:                             10 ft.
    Clay:                             20 ft.

*Water depth*
    Design values—
        Maximum:                      6,000 ft.
        Minimum:                      0 ft.

Experience—
    Maximum:                        6,000 ft.
    Minimum:                         30 ft.

*Limitations*

Relatively sensitive to wind, seas, and currents during installation, because of the relatively longer time during which the surface vessel must remain on station.
Fairly smooth and level seafloor required by the support-guidance frame.

*Advantageous features*

Expendable components of installation system are relatively inexpensive.
Penetration can be monitored and holding capacity predicted without prior investigation.
No lines to the surface other than anchor line.

3.12.3. *Details. Anchor assembly* (see Fig. 37).

| | |
|---|---|
| Height (based on 15-ft shaft): | 19 to 21 ft depending upon size of fluke. |
| Maximum diameter (support-guidance frame): | 8 ft. |
| Weight: | 1,800 lb. |

*Fluke-shaft assembly*

Fluke (see Fig. 38)—
    Type:                        Rotating "Y-fins" (three semi-circular steel plates joined along their straight edges to form a Y-section with 120-degree dihedral angles); upper half of one plate omitted to make room for keying linkage.

    Fluke diameter:           2, 3, and 4 ft.
    Plate thickness:         1/2 in.

Keying linkage—
    Type:                        Two-bar linkage between collar at base of shaft and outer corner of quarter-circle fin; fluke rotates when shaft moves upward.

Fluke-shaft locking mechanism (see Fig. 39)—
    Type:                        Two over-center toggles pinned to shaft, and tension straps from toggles to fluke.
    Function:                 Locks fluke securely to shaft during penetration; released by tripping slug at end of anchor cable inside shaft when upward load is applied to cable.

Shaft—
    Type:                        3-in. schedule 80 pipe.
    Length:                   15 ft (normal); readily varied.

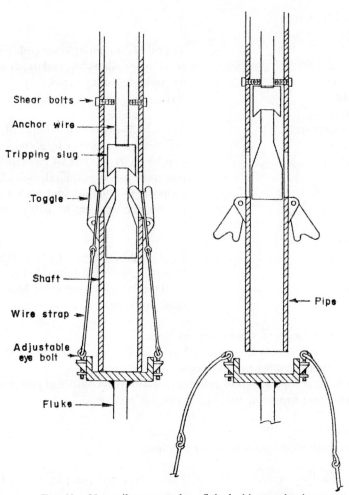

FIG. 39.   Navy vibratory anchor; fluke locking mechanism.

*Drive assembly*
    Vibrator—
        Type:                    Two counter-rotating masses.
        Location:              On shafts of motors in housing mounted on upper end of shaft.
        Peak force:           12,500 lb at 4,500 rpm.
    Motor—
        Number:             2.
    Type:                    Electrical (d.c.).
        Power:               4 hp.

*Support-guidance frame* (see Fig. 40).
    Base—
        Construction:       Welded hexagon of 3-in. pipe.

| | |
|---|---|
| Outside diameter: | 8 ft. |
| Support— | |
| Type: | Tripod of 3-in. pipe; lower ends pinned to base so as to be collapsible, and upper ends fastened to guide-sleeve segments. |
| Height: | 6 ft. |
| Guide sleeve— | |
| Construction: | Three 120-degree portions of a circular cylinder held together by a clamp. |
| Function: | Guides shaft at start of embedment process; proximity of vibrator releases clamp, allows supports to collapse, and permits penetration to continue until entire shaft is embedded. |

*Energy source*

| | |
|---|---|
| Type: | Lead–acid batteries (20 12-V, 30-amp-hr). |
| Life: | 60 min, approx (sustained load). |
| Location: | In three boxes mounted on base of support–guidance frame. |

3.12.4. *Operational aspects. Operational modes.*

1. Shallow water: cable-lowered, surface-powered, used without support guidance frame, drive assembly not recovered.
2. Deep water: Cable-lowered, automatic-starting, uncontrolled power supply, drive assembly and supports, guidance frame not recovered.

*Safety features*

Accommodates standard field safety practice.

3.12.5. *Cost.* Shallow water (<300 ft): $4,000 (approx) per placement. Deep water: $10,000 (approx) per placement.

3.12.6. *References*

1. Naval Civil Engineering Laboratory. Contract Report No. CR 69-009: Vibratory embedment anchor system. Long Beach, CA, Ocean Science and Engineering, Inc., Feb 1969. (Contract no. N62399-68-C-008) (AD848920L).
2. Naval Civil Engineering Laboratory. Technical Note N-1133: Specialized anchors for the deep sea; progress summary, by J. E. Smith, R. M. Beard, and R. J. Taylor. Port Hueneme, CA, Nov 1970. (AD716408).
3. Naval Civil Engineering Laboratory. Technical Report R-791: Direct embedment vibratory anchor, by R. M. Beard. Port Hueneme, CA, Jun 1973. (AD766103).

3.13. *Vibratory embedment anchor, Model* 2000 (*vibrated*).

3.13.1. *Source.* Ocean Science and Engineering, Inc., 5541 Nicholson Lane, Rockville, Maryland 20852, U.S.A.

3.13.2. *General characteristics.* An operational, low-cost, lightweight, fairly high capacity anchor for sediments. It is used in shallow-to-moderate depth (500 ft) for taut-line tethers, short-scope ship moorings, and other applications requiring precise placement of the anchor (see Fig. 41).

*Advertised nominal holding capacity*
    80,000 lb.

*Nominal penetration*
    40 ft.

*Water depth*
    Design values—
        Maximum:                  500 ft.
        Minimum:                   5 ft.
    Experience—
        Maximum:                  —
        Minimum:                   —

*Limitations*
    Relatively sensitive to wind, seas, and currents during installation, because of the relatively longer time during which the surface vessel must remain on station.
    Potential for entanglement of multiple lines.

*Advantageous features*
    Most of the installation equipment is recoverable and reusable.
    Penetration can be monitored and holding capacity predicted without prior investigation.

3.13.2. *Details. Anchor assembly* (see Fig. 41).
    Height (based on 40-ft shaft):     43 ft (approx).
    Maximum transverse dimension
        (tether bar):               7 ft.
    Weight:                     1,000 lb.

*Fluke assembly* (see Figs 41 and 42).
    Fluke—
        Type:                         Rotating "Y-fins" (three semi-circular steel plates joined along their straight edges to form a Y-section with 120-degree dihedral angles); upper half of one plate omitted to make room for keying linkage.
        Fluke diameter:           3 ft.
        Plate thickness:         3/8 in.
    Keying linkage—
        Type:                         Two-bar linkage between collar at base of

shaft and outer corner of quarter-circle fin; fluke rotates when shaft move upward.

*Shank assembly*
  Shaft—
    Construction:                4-in. schedule 40 pipe.
    Length:                      40 ft (normal; readily varied).
  Tension member—
    Construction:                3/4-in. wire rope inside the shaft, extending from the fluke to a tensioning device at the upper end of the shaft.
    Function:                    Secures the fluke, shaft, and drive assembly together during penetration.
  Tensioning device—
    Construction:                Hand-operated, 3,000-psi hydraulic cylinder by which collar on upper end of tension member is pulled upward.
    Location:                    Attached to vibrator housing.
  Tether bar—
    Construction:                Steel bar, 7 ft long, pin-connected to collar near upper end of shaft; collar swivels around shaft.
    Function:                    Point of attachment of anchor line; permits swinging of moored vessel.

*Drive assembly*
  Vibrator—
    Type:                        Two counter-rotating masses.
    Location:                    On shafts of motors in housing mounted on upper end of shaft.
    Peak force:                  24,000 to 30,000 lb at 3,600 rpm.
  Motor—
    Number:                      2.
    Type:                        Hydraulic.
    Capacity:                    17 gpm at 3,600 rpm.

*Power source*
  Pump—
    Type:                        Hydraulic, variable positive-displacement.
    Location:                    On surface vessel.
    Capacity:                    0 to 27 gpm.
    Pressure:                    0 to 3,000 psi.
  Prime mover—
    Type:                        Diesel engine.
    Location:                    On surface vessel.
    Size:                        6-cylinder, 100-hp.

### 3.13.4. *Operational aspects. Operational modes.*

Cable-lowered, remote–manual starting, remote–manual control of power supply drive assembly recovered.

*Safety features*

Accommodates standard field safety practice.

### 3.13.5. *Cost.* The material cost per anchor installation when purchased in lots of from 1 to 50 anchors is $3,184. This assumes one reusable drive assembly per 100 installations with the drive assembly costing $23,400.

### 3.13.6. *References*

1. S. H. Shaw. "New anchoring concept moors floating drydock", Ocean Industry, Vol 7, no. 1, Jan 1972, pp 31–33.
2. Letter, R. L. Fagan (OS&E) to R. J. Taylor (CEL), 11 Dec 1973.
3. Ocean Science and Engineering, Inc. Pamphlet: New anchoring system: Vibratory embedment anchor, Model 1000, Rockville, MD, undated.

### 3.14. *Chance special offshore multi-helix screw anchor (screw-in).*

### 3.14.1. *Source.* Anchoring Inc., P.O. Box 55263, Houston, Texas 77055, U.S.A.

### 3.14.2. *General characteristics.* An operational, reliable anchor for use primarily in sediments and in moderately shallow water. It can be installed rapidly and precisely, and it is used extensively for pipeline tiedown (see Fig. 42).

*Advertised nominal capacity*
    10,000 lb.

*Nominal penetration*
    10 ft.

*Water depth*
    Design values—
        Maximum:                —
        Minimum:                —
    Experience—
        Maximum:                325 ft.
        Minimum:                0 ft.

*Limitations*

For precise placement, relatively quiet conditions (wind, wave, current) are required during positioning of the drive system and initial phase of embedment.

*Advantageous features*

Several simple options are available for increasing holding capacity: diameter, number, and spacing of helixes; torsional strength of shaft; and depth of penetration.

### 3.14.3. *Details. Anchor* (see Fig. 43).

Type:
: Two to four circular, single-turn, helical surfaces spaced along a circular shaft (square shafts available).

Shaft outside diameter:
: 1-1/4 in. (larger sizes available).

Shaft length—

Anchor section (carries helixes):
: 5, 7, or 10 ft for two, three, or four helixes, respectively.

Extension section (no helixes):
: 10 ft maximum.

Helix diameter:
: 4 to 6 in. (larger sizes available).

Weight:
: 100 lb (average).

*Installation assembly* (*pipeline anchors*) (see Figs 42 and 44).

Size:
: Approx 3 × 5 × 8 ft without anchors.

Weight:
: 6,000 lb.

Drive heads—

Type:
: Two counter-rotating, gear-driven.

Speed:
: 45 rpm.

Maximum torque:
: Greater than 5,000 ft-lb.

Motors—

Number:
: 2.

Type:
: Hydraulic.

Flow rate:
: 25 to 30 gpm.

Buoyancy tank:
: —

*Power system*

Location:
: On support vessel.

Size:
: Approx 5 × 5 × 5 ft.

Pump—

Flow rate:
: 25 to 30 gpm.

Maximum pressure:
: 2,000 psi.

Prime mover:
: Diesel engine.

Air compressor:
: —

### 3.14.4. *Operational aspects. Operational modes* (*pipeline anchors*).

1. Cable-lowered, position controlled by divers, installation assembly recovered.
2. Cable-lowered, position controlled by television, installation assembly recovered.

*Safety features*

Accommodates standard field safety practice.

### 3.14.5. *Cost.* Approximately $375 per pair, installed. Anchoring Inc. installs all the anchors.

FIG. 40.   Navy vibratory anchor; embedded in sand on beach to demonstrate collapsible support-guidance frame.

FIG. 41. Ocean Science and Engineering vibratory embedment anchor, Model 2000.

FIG. 42.   Chance Special Offshore Multi-Helix system for pipeline anchoring; pipeline bracket
visible.

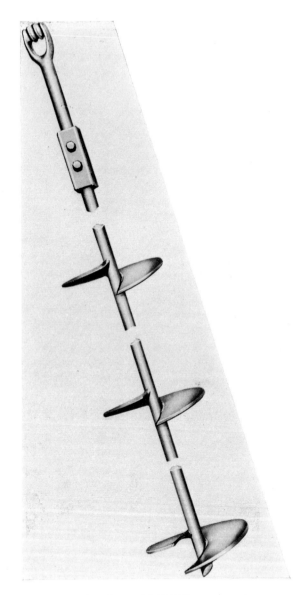

FIG. 43.  Chance Multi-Helix screw anchor.

FIG. 44.   Chance Special Offshore Multi-Helix system for pipeline anchoring.

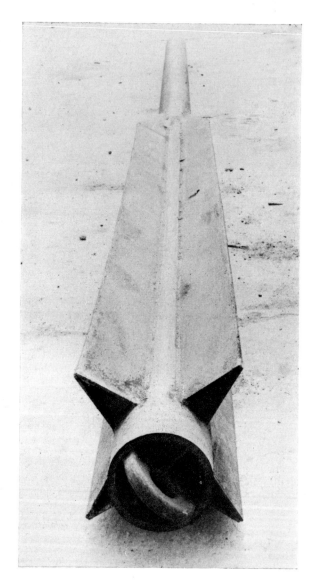

FIG. 45.   Navy stake pile; 8-in.

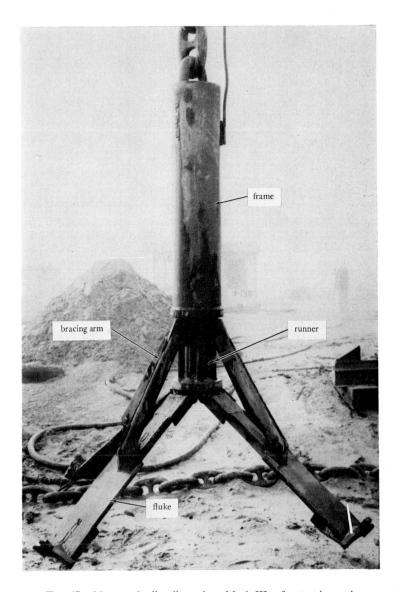

FIG. 47.   Navy umbrella pile-anchor, Mark III; after test in sand.

FIG. 50. Navy umbrella pile-anchor, Mark IV; after test in sand.

### 3.14.6. *References*

1. A. B. Chance Co. Bulletin 424-C: No-wrench screw anchors, Centralia, MO, 1969. (Part C of Encyclopedia of Anchoring).
2. G. E. Cannon. "Pipe anchors pin line solidly to sea floor", Offshore, Vol 29, no. 12, Nov 1969, pp 84, 86.
3. Letter, G. E. Cannon (Anchoring, Inc.) to R. J. Taylor (CEL), Sep 4, 1973.

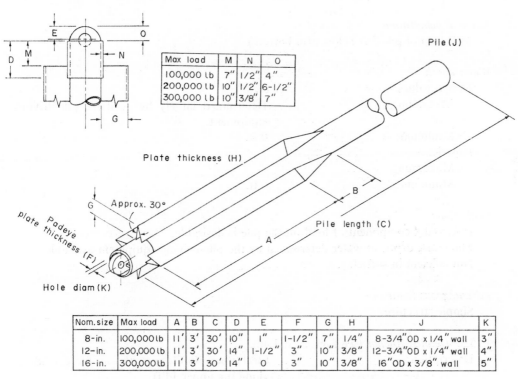

| Max load | M | N | O |
|---|---|---|---|
| 100,000 lb | 7″ | 1/2″ | 4″ |
| 200,000 lb | 10″ | 1/2″ | 6-1/2″ |
| 300,000 lb | 10″ | 3/8″ | 7″ |

| Nom. size | Max load | A | B | C | D | E | F | G | H | J | K |
|---|---|---|---|---|---|---|---|---|---|---|---|
| 8-in. | 100,000 lb | 11′ | 3′ | 30′ | 10″ | 1″ | 1-1/2″ | 7″ | 1/4″ | 8-3/4″OD x 1/4″ wall | 3″ |
| 12-in. | 200,000 lb | 11′ | 3′ | 30′ | 14″ | 1-1/2″ | 3″ | 10″ | 3/8″ | 12-3/4″OD x 1/4″ wall | 4″ |
| 16-in. | 300,000 lb | 11′ | 3′ | 30′ | 14″ | 0 | 3″ | 10″ | 3/8″ | 16″OD x 3/8″ wall | 5″ |

FIG. 46.   Design specifications for Navy stake pile.

### 3.15. *Stake pile* (*driven*).

3.15.1. *Source.* Naval Facilities Engineering Command, 200 Stovall Street, Alexandria, Virginia 22332, U.S.A.

3.15.2. *General characteristics.* An operational anchor that has been tested and used in East Coast locations to secure mothball fleets. It comprises a family of moderate-to-large-capacity anchors for Fleet moorings for ships and floating drydocks that will not drag under load and does not require dragging for setting.

*Advertised nominal holding capaciiy*
    For 8-in. pile size—
        Sand:                    100,000 lb.
        Soft clay:            20,000 lb.

For 12-in. pile size—
    Sand:                              200,000 lb.
    Soft clay:                      30,000 lb.
For 16-in. pile size—
    Sand:                              300,000 lb.
    Soft clay:                      40,000 lb.

*Nominal penetration*
35 ft (top of pile 5 ft below firm bottom).

*Water depth*
    Design values—
        Maximum:               Determined largely by available pile-driving equipment.
        Minimum:               0 ft.
    Experience—
        Maximum:               —
        Minimum:               0 ft.

*Limitations*
Horizontal component of load on the pile is desirable.
Maximum depth of water determined by the pile-driving equipment available.
Not efficient in soft clay.

*Advantageous features*
Simple structure.

3.15.3. *Details. Anchor* (see Figs 45 and 46).
    Description:              Steel tubes, 30 ft long, with four fins extending along the upper 14 ft.

Outside diameter—
    For 8-in. pipe:           8.75 in.
    For 12-in. pipe:        12.75 in.
    For 16-in. pipe:        16.00 in.
Pipe wall thickness—
    For 8-in. pipe:           0.25 in.
    For 12-in. pipe:        0.25 in.
    For 16-in. pipe:        0.375 in.
Width of fins—
    For 8-in. pipe:           7 in.
    For 12-in. pipe:        10 in.
    For 16-in. pipe:        10 in.
Weight—
    For 8-in. pipe:           1,400 lb.
    For 12-in. pipe:        2,600 lb.
    For 16-in. pipe:        3,600 lb.

3.15.4. *Operational aspects. Operational modes.*
1. Surface driving.
2. Underwater driving.

*Safety features*
Accommodates standard field safety practice.

3.15.5. *Cost.* For 8in. size: $2,500 ea (approx). For 12-in. size: $3,100 ea (approx). For 16-in. size: $3,600 ea (approx). Costs are for hardware only; offshore pile driving currently costs $10,000 to $15,000 per day.

3.15.6. *References*
1. Naval Civil Engineering Laboratory. Technical Note N-205: Stake pile development for moorings in sand bottoms, by J. E. Smith. Port Hueneme, CA, Nov 1954. (AD81261).
2. Naval Civil Engineering Laboratory. Letter Report L-022: Stake pile tests in mud bottom, by J. E. Smith. Port Hueneme, CA, Sep 1957.

3.16. *Umbrella pile-anchor, Mark III (driven).*

3.16.1. *Source.* Naval Facilities Engineering Command, 200 Stovall Street, Alexandria, Virginia 22332, U.S.A.

3.16.2. *General characteristics.* This item has been tested, but not used. It is an anchor for moorings for large vessels which (1) will not drag under load, (2) does not require dragging for pre-setting, and (3) has high capacity in bearing and in resistance to uplift.

*Advertised nominal holding capacity*
Sand:                                                        300,000 lb.

*Nominal penetration*
20 ft.

*Water depth*
Design values—
Maximum:                              Determined largely by available pile-driving equipment.
Minimum:                               0 ft.
Experience—
Maximum:                              Not tested offshore.
Minimum:                               0 ft.

*Limitations*
Maximum depth of water determined by the pile-driving equipment available.
Use restricted to homogeneous, uncemented soils free of boulders and other obstructions.
Design not well-adapted to development of a family of anchors of varying size.

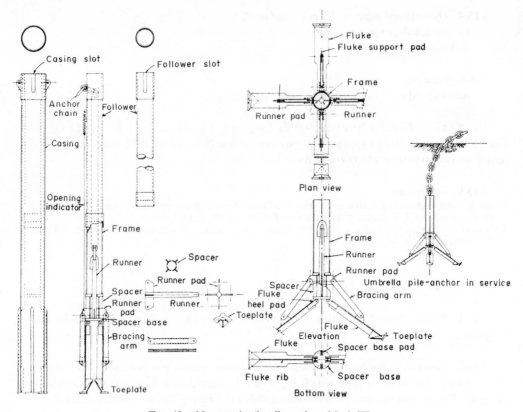

FIG. 48. Navy umbrella pile-anchor, Mark III.

*Advantage features*

Large capacity in both bearing and resistance to uplift in sand and cohesive soil.

3.16.3. *Details. Fluke assembly* (see Figs 47 and 48).

| | |
|---|---|
| Type: | Expanding finger type (four flukes). |
| Length of fluke: | 52 in. |
| Width of fluke: | 10 in. |
| Effective area of flukes: | 10.5 ft². |
| Angle of rotation from fully closed to fully opened positions: | 60 deg. |
| Outside diameter (foot circle of open flukes): | 8 ft. |
| Height of assembly: | 10 ft. |
| Weight of assembly: | 1,400 lb. |

*Chain*

| | |
|---|---|
| Size: | 2-3/4-in. |
| Length: | See length of follower. |

*Follower*
| | |
|---|---|
| Construction: | Steel tubing. |
| Outside diameter: | 12.75 in. |
| Length: | Varies with water depth and embedment depth. |

*Casing*
| | |
|---|---|
| Construction: | Steel tubing. |
| Outside diameter: | 18.0 in. |
| Length: | Varies with water depth and embedment depth. |

3.16.4. *Operational aspects. Operational modes.*
1. Surface driving.
2. Underwater driving.

*Safety features*
    Accommodates standard field safety practice.

3.16.5. *Cost.* $4,500 (approx) per anchor unit. Cost is for hardware only; offshore pile driving currently costs $10,000 to $15,000 per day.

3.16.6. *References*
1. Naval Civil Engineering Laboratory. Technical Report R-247: Umbrella pile-anchors, by J. E. Smith. Port Hueneme, CA, May 1963. (AD408404).

3.17. *Umbrella pile-anchor, Mark IV (driven).*

3.17.1. *Source.* Naval Facilities Engineering Command, 200 Stovall Street, Alexandria, Virginia 22332, U.S.A.

3.17.2. *General characteristics.* This item has been tested but not used. It is an anchor for moorings for large vessels which (1) will not drag under load, (2) does not require dragging for pre-setting, and (3) has high capacity in bearing and in resistance to uplift.

*Advertised nominal holding capacity*
| | |
|---|---|
| Sand: | 300,000 lb. |
| Mud: | 100,000 lb. |

*Nominal penetration*
    20 ft.

*Water depth*
    Design values—
| | |
|---|---|
|     Maximum: | Determined largely by available pile-driving equipment. |

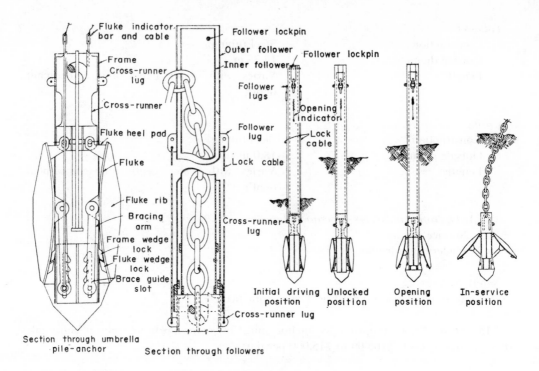

FIG. 49.   Navy umbrella pile-anchor, Mark IV.

|  | |
|---|---|
| Minimum: | 0 ft. |
| Experience— | |
| Maximum: | 35 ft. |
| Minimum: | 0 ft. |

*Limitations*

Maximum depth of water determined by the pile-driving equipment available.

Use restricted to homogeneous, uncemented soils free of boulders and other obstructions.

*Advantageous features*

Large capacity in both bearing and resistance to uplift.

Functional in sand and cohesive sediments.

3.17.3.   *Details. Fluke assembly* (see Figs 49 and 50).

| | |
|---|---|
| Type: | Expanding finger type (four flukes). |
| Length of fluke: | 49 in. |
| Width of fluke: | 14 in. |
| Effective area of fluke: | 16.5 ft². |
| Angle of rotation from fully closed to fully opened positions: | 60 degrees. |

Outside diameter (foot circle of
   open flukes):                          8 ft.

Height of assembly:                  8 ft.

Weight of assembly:                  2,200 lb.

*Inner follower*

Construction:                      Steel tubing.

Outside diameter:                12.75 in.

Length:                           Varies with water depth and embedment depth.

*Outer follower*

Construction:                      Steel tubing.

Outside diameter:                16.0 in.

Length:                           Varies with water depth and embedment depth.

*Chain*

Size:                               2-3/4 in.

Length:                         See length of inner follower.

**3.17.4.** *Operational aspects. Operational modes.*
1. Surface driving.
2. Underwater driving.

*Safety features*

Accommodates standard field safety practice.

**3.17.5.** *Cost.* $7,500 (approx) per anchor unit. Cost is for hardware only; offshore pile-driving currently costs $10,000 to $15,000 per day.

**3.17.6.** *References*

1. Naval Civil Engineering Laboratory. Technical Report R-247: Umbrella pile-anchors, by J. E. Smith. Port Hueneme, CA, May 1963. (AD408404).

**3.18.** *Rotating plate anchor (driven).*

**3.18.1.** *Source.* Techniques Louis Menard, Centre d'Etudes Geotechniques, Boite Postale No. 2, 91 Longjumeau, France.

**3.18.2.** *General characteristics.* An operational, high-capacity embedment anchor in sediments for single-point moorings, anchoring in offshore oil operations, and other applications.

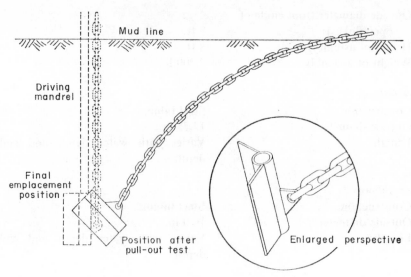

FIG. 51.   Ménard rotating plate anchor.

*Advertised nominal holding capacity*
    200,000 lb.

*Nominal penetration*
    10 ft to 30 ft.

*Water depth*
    Design values—
        Maximum:                          —
        Minimum:                          —
    Experience—
        Maximum:                          —
        Minimum:                          —

*Limitations*
    Maximum depth of water determined by the pile-driving equipment available.

*Advantageous features*
    —

3.18.3. *Details.*
*Fluke* (see Fig. 51).
    —

*Driving mandrel*
    —

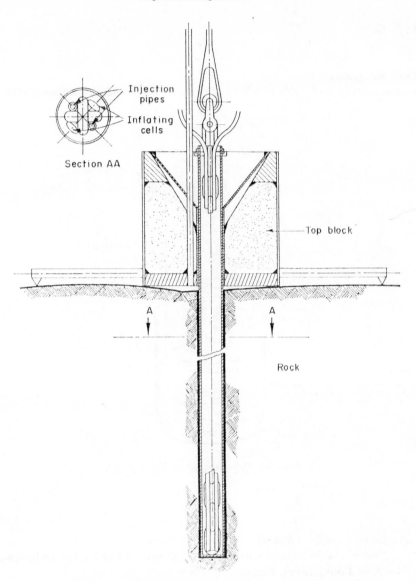

Fig. 52.  Ménard expanded rock anchor; placement of chain into drilled hole.

*Chain*

—

3.18.4. *Operational aspects. Operational modes.*
  1. Surface driving.
  2. Underwater driving.

*Safety features*

—

### 3.18.5. *Cost.*

—

### 3.18.6. *References*
1. Techniques Louis Menard. Publication P/95: Mooring Anchors. Longjumeau, France, 1970.

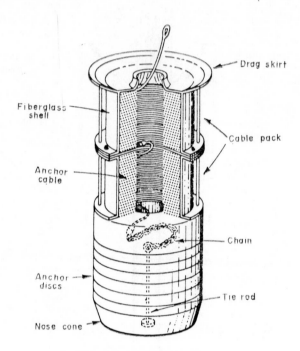

FIG. 54.   Delco free-fall anchor (typical).

## 3.19 *Expanded rock anchor* (*drilled*)

3.19.1. *Source.* Techniques Louis Menard, Centre d'Etudes Geotechniques, Boite Postale No. 2, 91 Longjumeau, France.

3.19.2 *General characteristics.* An operational, high-capacity anchor in rock for single-point moorings, anchorings for offshore oil operations, and other applications (see Fig. 52).

*Advertised nominal holding capacity*
   800,000 lb

*Nominal penetration*
   Rock:                                              30 ft.

*Water depth*                                     0 to 700 ft.

*Limitations*
Relatively long installation time.

*Advantageous features*
—

3.19.3 *Details.*
—

3.19.4 *Operational aspects.*
—

3.19.5 *Cost.*
—

3.19.6 *References.*
1. Techniques Louis Menard. Publication P/95: Mooring Anchors. Longjumeau, France, 1970,

3.20. *Free-fall anchor system (deadweight).*

3.20.1. *Source.* Delco Electronics, General Motors Corporation, 6767 Hollister Avenue, Goleta, California 93017, U.S.A.

3.20.2. *General characteristics.* An operational item used in numerous moorings (small ships, barges, buoys) in a wide range of depths. It minimizes the time, handling, and equipment required for installation.

*Advertised nominal holding capacity*
No fixed value. Anchor is usually custom-built, and size is readily varied over a wide range. Usual range is 600 lb to 24,000 lb (weight in air).
Resistance to uplift is approximately 85% of weight in air.
Resistance to horizontal force is variable, nominally 20% to 200% of weight in air depending upon seafloor.

*Nominal penetration*
Hard seafloor:                 0 ft.
Sediments:                     Variable, depending upon soil properties, water depth, anchor size.

*Water depth*
Design values—
  Maximum:                     20,000 ft.
  Minimum:                     100 ft (approx, for largest anchor).
Experience—
  Maximum:                     20,000 ft.
  Minimum:                     50 ft.

*Limitations*
Very heavy anchors in great depths are not retrievable with anchor line.

*Advantageous features*
Installation time minimized through elimination of on-station ship operations, such
as embedment or setting of anchor.
Deployable in relatively rough water.

3.20.3. *Details. Anchor assembly* (see Figs. 53 and 54).
Drag skirt diameter—
For 18-in.-OD size:              —
For 30-in.-OD size:              —
For 40-in.-OD size:              56 in.
Minimum height of assembly—
For 18-in.-OD size:              2-1/2 ft.
For 30-in.-OD size:              3 ft.
For 40-in.-OD size:              5 ft.
Maximum height of assembly—
For 18-in.-OD size:              4-1/2 ft.
For 30-in.-OD size:              6 ft.
For 40-in.-OD size:              13 ft.
Minimum weight of assembly—
For 18-in.-OD size:              600 lb.
For 30-in.-OD size:              1,400 lb.
For 40-in.-OD size:              4,000 lb.
Maximum weight of assembly—
For 18-in.-OD size:              3,000 lb.
For 30-in.-OD size:              6,000 lb.
For 40-in.-OD size:              24,000 lb.

*Nose cone*
Thickness—
For 18-in.-OD size:              —
For 30-in.-OD size:              —
For 40-in.-OD size:              —
Weight—
For 18-in.-OD size:              400 lb.
For 30-in.-OD size:              900 lb.
For 40-in.-OD size:              2,000 lb.

*Wafers*
Thickness—
For 18-in.-OD size:              —
For 30-in.-OD size:              —
For 40-in.-OD size:              3 in.

Weight—

|  |  |
|---|---|
| For 18-in.-OD size: | 200 lb. |
| For 30-in.-OD size: | 500 lb. |
| For 40-in.-OD size: | 1,000 lb. |

*Cable pack*
Maximum weight of cable—

|  |  |
|---|---|
| Any size: | 6,000 lb. |

*Cable (wire rope)*
Maximum size—

|  |  |
|---|---|
| For 1 × 19 stranding: | 3/8 in. |
| For 3 × 19 stranding: | 1-1/8 in. |
| For 3 × 46 stranding: | 1-1/2 in. |
| For 6 × 19 stranding: | 1-1/2 in. |

*Chain*

|  |  |
|---|---|
| Length: | 25 ft |
| Size: | 5/8 in. |

3.20.4. *Operational aspects* (see Figs 55 and 56). *Operational modes.*
Free-fall installation with cable deployed from cable pack(s) on the anchor.

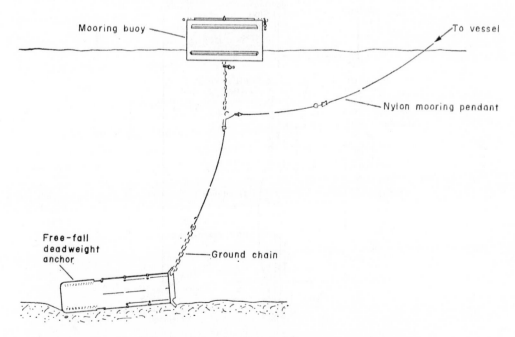

FIG. 56.   Delco free-fall anchor in typical deep-water mooring system.

*Safety features*

Accommodates standard field safety practice.

3.20.5. *Cost*. The cost per anchor ranges from $600 for a 600-lb anchor to $30,000 for a 24,000-lb anchor.

3.20.6. *References*

1. AC Electronics, Defense Research Laboratories. Manual No. OM69-01: Technical manual for Project BOMEX free-fall anchor systems. Santa Barbara, CA, Feb 1969. (Contract no. E-118-69(N)).
2. Delco Electronics. Report No. TR71-05: Containerized cable stowage, by J. Melendez. Santa Barbara, CA, Mar 1971. (Contract no. N00024-70-C-5474).
3. Letter, C. D. Leedham (Delco) to R. J. Taylor (CEL), Sep 20, 1973.

## 4. OTHER PROSPECTIVE TYPES

This section presents anchors that are still in the conceptual phase or initial phase of development or whose development were terminated due to technical problems. The Implosive, Free-fall, Pulse-Jet, Padlock, Jetted-In and Hydrostatic anchors and Seafloor Rock Fasteners are considered.

### 4.1 *Implosive anchor*

4.1.1. *Background*. The implosive anchor concept has only recently evolved. It utilizes hyd ostatic pressure as the energy source to embed a projectile into the seafloor. While the idea of the implosive anchor is new, the thought of using the abundant ocean energy to

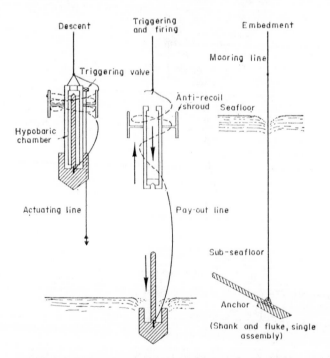

FIG. 57. Propelled-shaft embedment of implosive anchor (Rossfelder and Cheung, 1973).

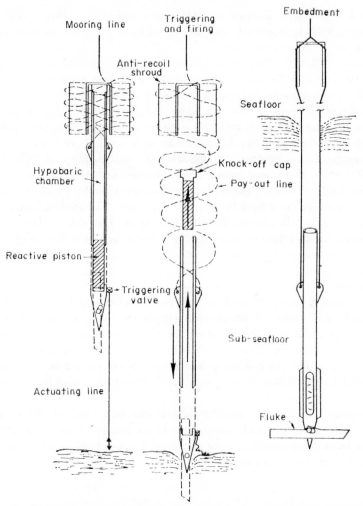

FIG. 58.  Propelled-casing embedment of implosive anchor (Rossfelder and Cheung, 1973).

perform useful work is not new. Dantz and Ciani (1967), who were concerned with developing power sources for the deep ocean, designed and built a single-impulse, hydrostatically powered ram device. The usefulness of this power source was verified. Frohlich and McNary (1969) designed and tested a hydrostatically actuated rock corer. They encountered some mechanical problems, but proved that the concept was feasible. The North American Rockwell Corporation actually fabricated an implosive anchor during the 1960s; however, information on this device could not be obtained.

4.1.2. *Description.* The implosive anchor (Rossfelder and Cheung, 1973) detailed in Figs 57 and 58 is similar in form to the propellant-actuated anchor in that it consists of two basic assemblies: the propelled part and the reactive part. The propelled part can either be mounted on an inner piston, which is displaced within a hypobaric breech by admission of the environmental pressure, or it can be the hypobaric chamber itself. The reactive-part can be fitted with a shroud to increase its added mass and limit its recoil, or for the case where the chamber itself is propelled, the reactive part can be either an inner piston with shaft and shroud or a free-inertial piston.

4.1.3. *Current status.* A feasibility study of the implosive anchor, which included development of a parametric model and performance of a parametric analysis, was conducted (Rossfelder and Cheung, 1973). The major findings were that: (1) anchor operation is influenced by chamber and environmental pressure differential, chamber volume, projectile mass and reactor effective mass, head losses at water entrance, and recoil losses; (2) piston and seals friction appear insignificant for design purposes; (3) for a given anchor mass at a given depth and with a given kinetic energy requirement, there is an optimum volume and geometric design of the hypobaric chamber; and (4) short stroke chambers appear more efficient than long stroke chambers. The study concludes that the concept is feasible and that the primary areas which remain to be addressed are design of the water admission device to minimize head loss, reactor design, and triggering mechanism design.

### 4.2 Free-fall anchor

4.2.1. *Background.* A "free-fall" anchor is one that falls freely to the seafloor and embeds through its own kinetic energy. Though holding capacities would be limited to moderate values, many urgent requirements for anchoring relatively small structures could be satisfied. Quick, easy, and more accurate placement of anchors could be achieved, and better holding power efficiency as measured by holding-power-to-weight ratio could be attained. Holding capacities of 15,000 to 25,000 lb were considered adequate values to meet these requirements.

4.2.2. *Description.* After minor modifications to the initial design, the CEL free-fall anchor, Fig. 59, evolved. It is a steel construction in the general shape of an arrow, and it consists of three basic components: a fluke assembly, a heavy steel shank, and a cable bale with protruding fins at the trailing end.

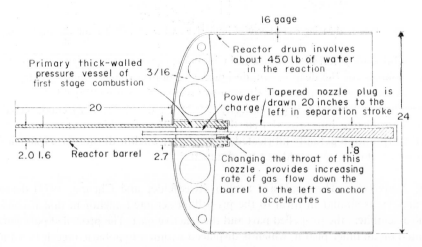

FIG. 60.   Mass drag reactor of the Pulse-Jet Anchor System (Lair, 1967).

FIG. 53. Delco free-fall anchor (12,000-pound anchor).

FIG. 55.  Delco free-fall anchor (24,000-pound); mounted on launching platform on USCGS Rockaway.

FIG. 59.    Free-fall embedment anchor.

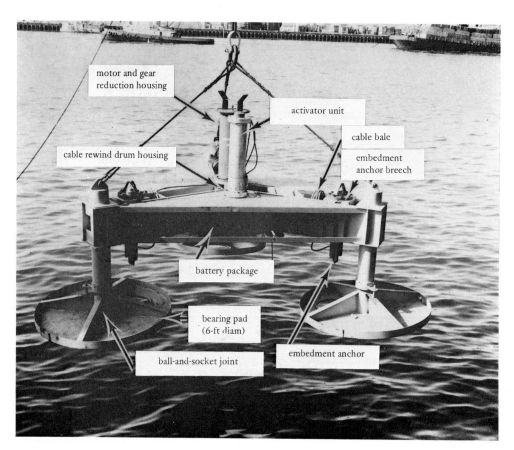

FIG. 63.  PADLOCK Anchor System developed for test and evaluation (Dantz, 1968).

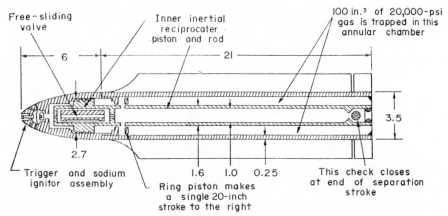

Fig. 61.   Ballistic embedding anchor of the Pulse-Jet Anchor System (Lair, 1967).

The anchor fluke is a special design which presents a minimum resistance to penetration and keys (rotates from the vertical to horizontal resistive position) rapidly to optimize use of anticipated limited penetration. The cable bale consists of cable coiled in a compact package; a reverse twist is placed in the cable for each coil. Without this reverse twist, the cable would tend to birdcage during pay-out, resulting in greatly reduced line strength and life.

4.2.3. *Current status.* As reported by Smith (1966) the free-fall anchor did not fulfil the requirement for being a practical, usable deep-sea anchor that could be free-dropped and, by its own impetus, embed into the seafloor and develop a holding capacity of sufficient amount to warrant its use in place of deadweights. The primary reason was that the size and configuration of the anchor necessary to accommodate the cable bale combined with the size and shape of the flukes necessary to obtain reasonable holding power was not compatible with attaining the velocity needed to obtain adequate embedment. For example, it was determined that even with the maximum theoretical velocity attainable by free-fall (about 35 fps), a holding-capacity-to-weight ratio of only 3 or 4 to 1 could be obtained. A minimum ratio of 7 to 1 is considered necessary for the free-fall anchor to be feasible.

Despite failure to achieve the idealized goal for a free-fall anchor, significant contributions toward development of improved, direct-embedment deep-sea anchors were realized. The cable pay-out system for deploying anchors in the deep sea works and has practical application within certain operational, size, and depth limitations. The knowledge and experience gained can be utilized in deploying future deep-sea anchors. More important is the revolutionary fluke incorporated into the design of the free-fall anchor. This fluke proved to be highly efficient and is adaptable to other types of direct-embedment anchors.

4.3. *Pulse-jet anchor*

4.3.1. *Background.* The concept for a pulse-jet anchor evolved during the investigation of explosive anchors at CEL. It became evident during testing of the explosive anchors that a power action extending throughout the embedment phase of anchor placement would more readily accommodate the variable resistance to penetration offered by seafloors comprised of firm and soft sediments. The pulse-jet principle could potentially achieve the

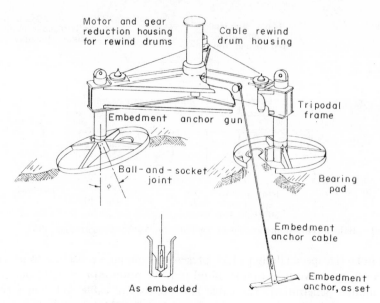

FIG. 62.   Basic concept of PADLOCK Anchor System (Dantz, 1968).

goal of extending the time during which power is applied to embed the anchor. The concept was investigated under contract by Sea Space Systems, Incorporated. The contractor was to design and fabricate two experimental models and to conduct developmental testing. Then two prototype models were to be delivered for Government testing.

The concept proved to be not feasible, and the contract was reduced in scope to include a report on the effort (Lair, 1967).

4.3.2. *Description.* The pulse-jet anchor as envisaged is comprised of two principal parts: a Mass Drag Reactor, Fig. 60, and a Ballastic Embedding Anchor, Fig. 61. The Ballistic Embedding Anchor is meshed with the Mass Drag Reactor, and the resulting assembly is lowered to the seafloor. On contact, a propellant in the Mass Drag Reactor gives the Ballistic Embedding Anchor an impetus to embed at least its own length into the seafloor. To this point, the principle is similar to that for other propellant-actuated anchors. The Ballistic Embedding Anchor consists of three main components: a main structural body, an inner inertial reciprocator that executes a short stroke with respect to the structural body, and an innermost free-sliding valve that executes a shorter stroke than the reciprocator and governs the stroke of the latter.

As the anchor is expelled from the Mass Drag Reactor, it traps and seals a charge of expulsion gases at about 20,000 psi. Beyond this point the principle differs from that of other explosive anchors. This charge of gas is distributed by the valve to drive the reciprocator up and down and ultimately is exhausted forward from the anchor nose to break up the seafloor in front of the advancing anchor. The embedment phase ceases when the gas pressure equals that of the ambient sea. Then a load is applied to the anchor to key it over to a position of maximum resistance.

4.3.3. *Current status.* The contractor was unable to achieve an experimental model of the design envisaged. Two ideas were reported as being too optimistic. The first related to the reciprocating machine in that sliding seals could not be made to function satisfactorily at the high temperatures and pressures encountered in the design. The second pertained to determining the critical relationship between the internal mechanics of the anchor and the soil mechanics of the seafloor. Extensive and expensive developmental testing was indicated for both problem areas with no assurance of success.

Two ideas were reported to have stood up under study and evaluation. The first was the concept of multiphase release of energy. The second was the forward jetting of exhaust gases to assist and regulate anchor embedment.

On review of the contractor's report, it was concluded that the cost to solve the problems for successful development of this concept was too great to warrant further investigation.

## 4.4. *PADLOCK anchor system*

4.4.1. *Background.* The PADLOCK anchor was designed to provide a high-capacity fixed-point (resistance to bearing, lateral, and uplift loads) anchoring system that could be installed without diver assistance. A feasibility program was initiated at CEL. The scope included the conception, design, fabrication, and evaluation of a self-contained anchor system that employs multiple bearing pads in conjunction with propellant-actuated anchors. The effort, currently suspended, was reported by Dantz (1968).

4.4.2. *Description.* The PADLOCK is a tripod framework constructed of lightweight materials and supported at each leg by articulated, round bearing pads. To obtain resistance to uplift, propellant-actuated direct-embedment anchors are incorporated into the system. The general scheme of the entire system is shown in Fig. 62. The bearing pads are connected to the frame with ball-joints that allow the pads to maintain maximum contact with the seafloor by adjusting to contour slopes as great as 10%. An anchor is housed above each of the bearings. After the anchors are propelled into the seafloor, they are set by pretensioning the embedment anchor cables with a rewind mechanism located in a central housing unit at the junction of the arms of the tripod framework. The objective is to clamp the pads to the seafloor by obtaining a firm hold in the seafloor soil with the anchors.

A propellant-actuated anchor was selected to develop the uplift resistance. The particular anchor design chosen was the Hove II (now VERTOHOLD) anchor. The commercial anchor of this style was rated as having a nominal 10-kip capacity, whereas a 20-kip capacity was desired. Therefore, the manufacturer had to build and deliver a specially enlarged size. The configuration, size, and load-supporting capacities selected were judged sufficient to demonstrate the feasibility of the system.

The PADLOCK prototype fabricated for testing and evaluation is shown in Fig. 63. A key feature of the concept is the cable rewind mechanism that pulls the anchors to a set position. The rewind mechanism consists of three separate cable drivers powered by a common shaft. Each drum holds the cable from one of the embedment anchors, and each could wind a sufficient length of cable to develop the pretension load for that anchor. Other features of the concept include: (1) an activator unit to control the sequence of operations of the PADLOCK by acoustic command once it is on the seafloor, (2) an ambient-pressure battery power source, and (3) a shipboard stern roller to assist in the installation of the PADLOCK.

4.4.3. *Current status.* Five shallow water tests were conducted with the PADLOCK in and about Port Hueneme Harbor in water depths from 18 to 60 ft. The seafloor was primarily hard-packed silty sand. In no single test did all of the components function as a complete system. However, each component performed separately as intended, at least once.

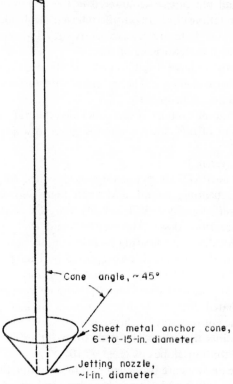

FIG. 64.   Illustration of Jetted Anchor.

Most of the problems involved the propellant-actuated anchors. For example, the contractor-procured anchors were found to be improperly heat-treated, and they failed under high acceleration-induced stress. This fault was corrected after two tests. The recoil of the anchor gun assembly was restricted by the tripod framework, thereby causing high stresses in the anchor and the framework. Problems were encountered with the cable pay-out system. The cable bale had to provide a sufficient amount of cable for the anchor, whose depth of penetration varied for each shot, and a means had to be provided for the rewind mechanism to draw off the remaining cable and develop a pretension in the line. A new frame was designed specifically to accommodate a workable cable payout system. The new structure then performed according to design.

The activator unit initially malfunctioned due to an intermittently operating transistor. After the trouble was remedied, the unit functioned according to design. The battery power source was initially used without a protective container (heavy grease provided insulation from seawater), and it was subject to deterioration. Later a battery container filled with transformer oil and covered with a flexible neoprene top to make the system pressure-compensated was used to prevent deterioration of the batteries.

Dantz (1968) concluded that:

"1.   In general, the PADLOCK Anchor System has been demonstrated to be a workable concept.

2. The power supply, rewind mechanism, and cable system are workable and fully dependable.

3. The activator unit is operational, water tight at pressures up to 500 psi (no upper limit established), and not affected by the shock loads imposed by the detonation of the embedment anchors.

4. According to a limited number of tests, the reliability of all the components functioning as a complete system is very low, mainly because the reliability of the embedment anchors was unsatisfactory."

In 1968 it was recommended that further effort be suspended until the reliability of propellant-actuated embedment anchors was improved.

### 4.5. *Jetted-in anchor*

4.5.1. *Background.* The jetted-in approach to anchor embedment can be and/or has been applied to a variety of anchor types, including piles, deadweights, mushroom anchors, and simplified cone anchors. These inexpensive, diver-emplaced jetted-in anchors are capable of sustaining low-to-moderate uplight loads (2 to 10 kips). These anchors would be used for pipe and cable tie-downs, instrument pack tiedowns, and pulling points for underwater construction.

This procedure is considered more applicable in sand seafloors due to the liquefaction potential of this medium. Limited experimental data are available on the increased capacities of large jetted-in anchors; however, there have been tests run on small diver-emplaced anchors that are pertinent to this handbook. This discussion pertains solely to small cone anchors as reported by Stevenson and Venezia (1970).

4.5.2. *Description.* The jetted-in anchor, Fig. 64, is a buried vertical pipe that is forced into the seabed by a jetting action. Water is pumped into the upper end of the pipe and discharged at the bottom, thereby dislodging soil and permitting the pipe to settle in the hole. An enlarged section, such as a cone-shaped shield at the bottom, and backfilling and grouting the hole are means for improving the holding capacity.

4.5.3. *Current status.* Twenty-three anchors were diver-emplaced in coral sand; holding capacities varied from 2 to 10 kips. The installation procedures were simple and posed no

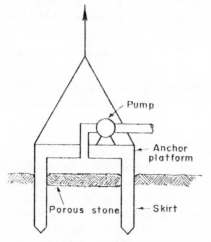

FIG. 65.   Schematic of the Hydrostatic Anchor.

problems to the divers. The grouting technique was very time consuming and needs refinement.

### 4.6 Hydrostatic anchor

4.6.1. *Background.* The need for an anchor that could provide short-term vertical resistance to breakout of submersibles and bottom resting platforms was evident. To satisfy this need, work was initiated at the University of Rhode Island on the development of a short-term high-efficiency anchor that utilized suction to develop its capacity (Brown and Nacci, 1971).

4.6.2. *Description.* The hydrostatic anchor, Fig. 65, is comprised of an anchor platform, a penetration skirt, a pump, a lifting harness, and a porous stone. The porous stone is necessary to prevent liquefaction of the soil beneath the stone.

4.6.3. *Current status.* According to Wang *et al.* (1974) the vertical breakout behavior

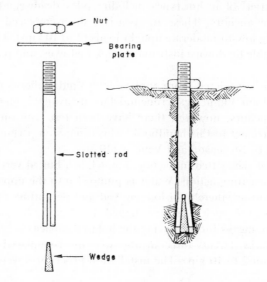

FIG. 66.   Drive-set rock bolt; slot and wedge type (Brackett and Parisi, 1975).

of the hydrostatic anchor depends greatly upon the anchor geometry (including anchor diameter and skirt length), soil strength properties, and the pressure difference between the ambient pressure and the pressure beneath the porous stone. The results of model tests indicate that the hydrostatic anchor functions most effectively in sand with decreasing effectiveness in silts and clays.

### 4.7. Seafloor rock fasteners

4.7.1. *Background.* Seafloor anchors available for shallow-water installation include a variety of seafloor rock fasteners, such as rock bolts, rebar, and drilled and grouted chain. Diver-installed fasteners have been used extensively to stabilize oceanographic cables, to secure structures to rock seafloors, and to moor small vessels. CEL has been attempting to improve the equipment and techniques for installing and, where applicable, grouting the fasteners to the seafloor (Brackett and Parisi, 1975; Parisi and Brackett, 1974).

TABLE 1.  PARAMETERS AFFECTING HOLDING STRENGTH OF SEAFLOOR FASTENERS

| Parameter | Effect on holding strength | Comments |
|---|---|---|
| Bolt diameter | The bolt diameter determines the ultimate potential holding strength possible for a given size bolt, and the ultimate tensile strength. | If all bolts have the same ultimate tensile strength, the failure load of the bolt will vary as the square of the diameter. |
| Anchor configuration<br>  Length and<br>  diameter of collar | The length and diameter of the anchor collar affect the stress produced in the seafloor rock. An increase in size of the anchor collar will decrease the stresses in the rock, thus reducing the chance of failure due to localized crushing or splitting of the rock. | An increase in anchor diameter requires an increase in drilling time. The trade-off between installing one large rock bolt or several small bolts in a padeye configuration should be considered. |
| Type of collar | A one-piece split collar has proven to give slightly higher pullout loads than the two-piece collar design for the same size fastener. | |
| Embedment depth | An increase in embedment depth produces almost a linear increase in holding strength up to the point where either localized crushing of the rock occurs around the collar or the ultimate tensile strength of the bolt is exceeded. | As a general rule a 6-inch embedment is sufficient to eliminate failure due to surface fracturing of the rock. Bolt diameter, competency of the rock, and presence of hard or soft substrata should be considered before determining the minimum embedment depth. |
| Duration of installation | There are not sufficient data at the present time to predict the exact effect of corrosion on the long-term holding strength of the fasteners tested. A trend toward a slightly reduced holding strength was detected after as little as 6 months of exposure. | The use of zinc anodes along with periodic inspection and replacement of spent anodes should ensure the integrity of the fastener for many years. |
| Initial torque | Initial torques of 40 ft-lb for the masonry stud anchor and 100 ft-lb for the spin-lock rock bolt were found to be necessary to properly set the anchor. Torquing the bolts above these values have no effect on the holding strength of the bolt. | The masonry stud anchors could be properly set by a diver using a hand wrench, but the use of an hydraulic impact wrench is recommended to ensure proper setting of the spin-lock rock bolt. |
| Compressive strength of rock | The holding strength of a given size fastener is almost linearly dependent on the unconfined compressive strength of the rock. | The presence of internal voids or fractures in the rock must be investigated before using compressive strength as a design criterion. |
| Installation of fasteners on land vs underwater | There appears to be a slight decrease in holding strength for bolts installed underwater compared with the same installation on land. The wide scatter of data points makes it difficult to quantitatively determine the magnitude of this decrease in failure load. However, if a normal safety factor is applied to the results of land tests, a realistic safe working load for the underwater installation should be obtained. | Care must be taken when using land tests to predict underwater performance. The test installations must be conducted in rock representative of that actually found at the seafloor work site. This analysis should include: size, porosity, presence of voids and fractures, presence of biological organisms, such as those in coral, that may have a significant effect on the holding strength of the fastener. |

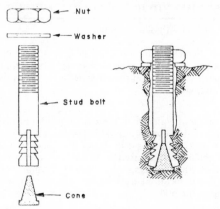

Fig. 67.   Drive-set rock bolt; cone and stud anchor type (Brackett and Parisi, 1975).

This section refers specifically to the rock bolt type of seafloor fastener and is generally derived from Brackett and Parisi (1975).

4.7.2. *Description.* Little data is available on grouted rock bolts; this section will be confined to the nongrouted type. All nongrouted rock bolts utilize the same principle to develop their anchoring strength. By mechanically expanding the down hole end of the bolt, an anchoring force is obtained through a combination of friction, adhesion between the anchor and rock, and physical penetration of the anchor into the rock. Rock bolts can generally be classified into two types: (1) drive-set, and (2) torque-set.

The slot and wedge bolt, Fig. 66, and cone and stud anchor, Fig. 67, are common examples of the drive-set type. The anchor is secured by placing the wedge into the slot and positioning the rod into the predrilled hole, then by driving the slotted rod over the wedge (which rests on the bottom of the hole) the rod expands into the rock.

Successful installation of the drive-set fastener depends on accurate hole drilling to a predetermined depth and the application of sufficient force to completely expand the slotted rod. Problems can also be encountered in soft rock where the driving force causes the wedge to be pushed into the rock rather than expanding the anchor.

A typical torque-set anchor is shown in Fig. 68. This type of rock anchor has a wedge or cone that is threaded to the bottom of the bolt. A sleeve or shell that surrounds the cone is pushed into the hole with the bolt. Once the bolt has been inserted, torque is applied to the nut to pull the bolt and cone up through the sleeve, thus securing the anchor.

The torque-set bolt requires far less precision in hole drilling providing the depth is greater than the length of the bolt. Expansion of the anchor is also unaffected by the quality of the rock at the bottom of the hole.

With the hand-held and hydraulically powered tools currently available to the under-water construction and salvage divers, it is easier to provide the torque for installing the torque-set type of anchor than the linear impact for installing the drive-set type.

4.7.3. *Current status.* Table 1 summarizes the parameters affecting the performance of seafloor rock bolts.

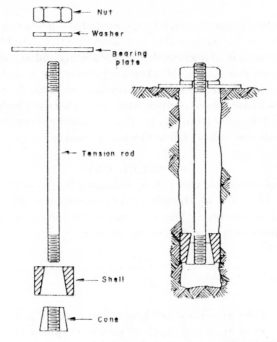

FIG. 68.   Torque-set rock bolt (typical) (Brackett and Parisi, 1975).

Work to date on diver-installed grouted fasteners has primarily involved development of a grout-dispensing device. The device is workable but must be lightened prior to Fleet usage. Testing on grouted fasteners has been minimal, but results indicate that the rock bolt type of fastener is superior to grouted fasteners because it is far simpler and quicker to install.

## 5. APPLICABLE COMPUTATIONS

The determination of the holding capacity of anchors designed to resist uplift loads involves considerations and techniques not required for conventional anchors. Conventional anchors are designed to embed as they are dragged. Should applied loads exceed their capacity, they will displace laterally but generally will continue to maintain their approximate design holding capacity once the excess loading eases. However, uplift-resisting anchors must be embedded by some means other than the service loading force, such as by drilling, driving, or ballistic propulsion. Once the uplift-resisting anchor is at its deepest penetration achieved during installation, all subsequent in-service applied loads will tend to extract it. Slight initial upward movements tend to seat it and mobilize the surrounding soil medium to resist extraction. Any excess loading on and/or movement of the anchor causes a reduction of the rated capacity and eventually causes extraction.

It is evident then that determining the penetration of an uplift-resisting anchor is important. Also, determining the initial movements to mobilize the soil and, in the case of anchors with outward folding flukes, determining the fluke-keying distance are important. Thus, in the next section computations to determine penetrations and fluke-keying distances are considered. Then in the following section methods for predicting holding capacities are presented.

### 5.1. *Penetration*

Penetration depths cannot be analytically predicted reliably in coral and rock. The soils are separated into two categories: clay and sand. Analytical techniques are provided for estimating the penetration of anchors driven ballistically and by vibration.

5.1.1. *Momentum penetration.* Momentum penetration is defined as the penetration achieved from its own momentum. The momentum can result from the anchor being fired from a gun, in which case the fluke or projectile is travelling at a high velocity when it strikes the seafloor. Or it can result from its own free-fall impetus. Propellant-actuated, implosive, and free-fall embedment anchors fall in this penetration category.

#### 5.1.1.1. *Clay*

Momentum penetration in clay can be estimated by the methods established by True (to be published). Equations for the solution of penetration problems are not suitable for a closed-form solution. However, they can readily be solved by incremental techniques. The incremental form to be used for computations is:

$$v_{i+1} = v_{i-1} + \frac{W - F_i(v_{ii} \, z_i)}{M^* \, v_i} (2 \, \Delta z) \tag{5-1}$$

where $v_{i+1}$ = Velocity at the depth being considered (ft/sec).

$v_{i-1}$ = Velocity at two depth measurements above the depth being considered (ft/sec).

$W$ = Buoyant weight of projectile in soil (lb).

$v_i$ = Velocity calculated one depth increment above the depth being considered (ft/sec).

$F_i(v_i, z_i)$ = Resisting force at the depth and velocity one depth increment above the depth being considered = $F_i^* + F_{Hi}$ (lb).

TABLE 2.  VALUES OF SIDE ADHESION FACTOR, $\delta^*$, AT HIGH
VELOCITY DERIVED FROM FIELD TEST DATA

| Projectile shape | Slenderness ratio, $l/D$ | High-velocity side adhesion factor, $\delta^*$ |
|---|---|---|
| Stubby | 9 | 0.11 |
| Medium | 15 | 0.23 |
| Slender | 30 | 0.46 |

$M^*$ = Effective mass of penetrator; equals penetrator mass plus added mass (slug).

$\Delta z$ = Depth increment (ft).

$F_i^*$ = Soil resisting force = $C_{1i} \, S_{ui} \, S_{ei}$ (lb).

$F_{Hi}$ = Fuid inertial drag force = $v_i^2 \, C_2$ (lb).

$S_{ui}$ = Undrained sediment shear strength (psf).

$S_{ei}$ = Ratio between dynamic and static shear strength

  = $S_{ei}^*/1 + [1/\sqrt{(C_e \, v_i/S_{ui} \, l_i)} + C_o]$.

$C_{1i}$ = $N_c A_{Fi} + (\delta_i^*/S_{ut}) \, A_{si}$ (ft$^2$).

$C_{2i} = (1/2)\rho_i\, C_D\, A_{Fi}.$

$S_e^* = $ Maximum $S_{ei}$ at high velocity; equal to 5 for all soils.

$C_e = $ Constant; equal to 20 for all clays and sands (psf-sec).

$I_i = $ Effective length of shearing zone; equals depth of embedment or length of penetrometer body, whichever is smaller (ft).

$C_o = $ Dimensionless constant; equal to 0.04 for all clays and sands.

$N_{\ni} = $ Deep bearing factor; equal to 9 for clays and sands.

$A_{Fi} = $ Frontal area of penetrometer (ft²).

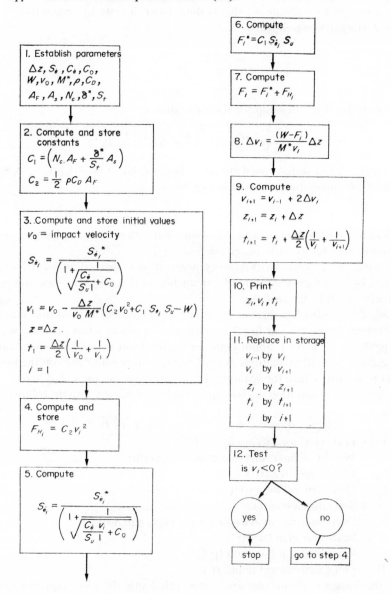

FIG. 69.   Incremental calculation flow for momentum penetration in clay and sand.

$\delta_i^*$ = Adhesion reduction factor (see Table 2).

$S_{ti}$ = Soil sensitivity (ratio of remoulded to undisturbed strength); use $S_{ti}$ = 1 for sands.

$A_{Si}$ = Side area of penetrometer (ft²).

$\rho_i$ = Mass density of soil (slug/ft³).

$C_D$ = Drag coefficient (estimated from fluid mechanics principles).

In Equation 5-1, all functions are known except $v_{i+1}$; $\Delta z$ is specified at one-twentieth or less of an estimated embedment depth. When beginning, however, $v_i = v_1$ is not known, and it is necessary to estimate $v_1$; this is done most directly by computing $v_2$ for $v_1 = v_o$ and then starting over again using

$$v_1 = \frac{v_o + v_2}{2}.$$

An equivalent direct relationship for this procedure is

$$v_1 = v_o - \frac{\Delta_z}{v_o\,M^*}\,(C_{21}\,v_o{}^2 + C_{11}\,S_{e1}\,S_{u1} - W). \tag{5-2}$$

A better estimate of an initial $v_1$ will not give a better value of final depth, $z_n$. A flow diagram of the calculation procedure is shown in Fig. 69.

### 5.1.1.2. Sand

Momentum penetration in sand can be estimated with the same techniques and equations used for estimating momentum penetration in clay.

5.1.2. *Vibratory penetration.* Vibratory penetration is defined as penetration gained by transmitting high-frequency vibration to an anchor so that under its own and/or additional weight it will sink into the seafloor.

Schmid (1969), who has discussed vibratory penetration in sand and clay soil, states that a vibratory driver will fail to advance the driven object when the total weight of bias plus the peak driving force is about equal to the total soil resistance to penetration. Beard (1973) presented Schmid's equations as applied to anchors with flukes on the lower end of a long shaft to be driven into the seafloor. These equations are presented here.

*Clay.* Vibratory penetration in clay can be readily calculated with the following equation:

$$Q + \text{Bias} = A_{fs}\,c + A_{ff}\,N_c\,c + a_s\,c_r\,D \tag{5-3}$$

where    $Q$ = Peak vibrator driving force (lb).

Bias = Weight of fluke-shaft vibrator system (lb).

$A_{fs}$ = Fluke side area (ft²).

$A_{ff}$ = Fluke frontal area (ft²).

$c$ = Soil cohesion (psf).

$N_c$ = Deep bearing capacity factor for clay; equal to 9.

$a_s$ = Shaft unit area (ft²/ft).

$c_r$ = Remolded soil cohesion (psf).

$D$ = Fluke embedment depth (ft).

For clays that have a uniform cohesion profile with depth, the above equation can be solved directly for the embedment depth. When the cohesion profile varies as a complex function of depth, it is necessary to solve the equation by trial and error because a particular cohesion value implies a particular depth. However, for seafloor soils the cohesion profile is often specified by a constant function of depth in the form of a ratio of cohesion to effective overburden pressure. Multiplying this ratio by depth and buoyant soil density gives the cohesion at that depth. (The remolded cohesion is attained by dividing the cohesion by the soil sensitivity.) When this is the case, Equation 5-3 becomes

$$Q + \text{Bias} = A_{fs} \frac{c}{p} \gamma_b\, D + A_{ff}\, N_c \frac{c}{p} \gamma_b\, D + a_s \left(\frac{1}{S_t}\right)\left(\frac{c}{p}\right) \gamma_b \frac{D^2}{2}. \tag{5-4}$$

This equation can be solved for depth in terms of the other parameters using the quadratic equation. The result is:

$$D = \frac{-(X + Y) \pm [(X + Y)^2 + 4\, W(Q + \text{Bias})]^{1/2}}{2\, W} \tag{5-5}$$

where  $X = A_{fs}\,(c/p)\gamma_b.$
  $Y = A_{ff}\, N_c\,(c/p)\gamma_b.$
  $W = (1/2)\, a_s\,(1/S_t)\,(c/p)\gamma_b{}^{/2}.$
  $c/p$ = Ratio of cohesion to effective overburden pressure.
  $\gamma_b$ = Buoyant unit weight of soil (pcf).
  $S_t$ = Soil sensitivity.

*Sand.* For sand the equation for vibratory penetration is

$$Q + \text{Bias} = A_{fs}\, \sigma_v\, K \tan \varphi_s + A_{ff}\, N_q\, \sigma_v + a_s\, \sigma_v\, K \tan\varphi_s\, \frac{D}{2} \tag{5-6}$$

where  $\sigma_v$ = Effective vertical pressure (psf).
  $K$ = Tatio of principal soil stresses.
  $\varphi_s$ = Friction angle between object and sand (deg.).
  $N_q$ = Deep bearing capacity factor for sand.

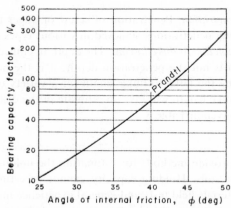

FIG. 70.  Theoretical capacity factor, $N_q$, vs angle of internal friction, $\varphi$, for a strip foundation.

It is recommended that $N_q$ values be chosen according to the curve in Fig. 70. Values of $K$ can be taken as 1.5 for dense sand and 1.0 for loose sand. The angle of friction between sand and smooth metal surfaces is independent of soil density and is taken as 26 degrees. For rough surfaces $\varphi_s$ should be taken as the angle of internal friction of the sand. When the density of the sand varies significantly with depth, Equation 5-6 must be solved by trial and error. If the sand has a uniform density over the depth of interest or if it can be approximated as such, Equation 5-6 can be rewritten by substituting the product of soil depth and soil buoyant density for the effective vertical pressure. Equation 5-6 then becomes:

$$Q + \text{Bias} = A_{fs}\, \gamma_b\, D\, K \tan \varphi_s + A_{ff}\, N_q\, \gamma_b\, D + a_s\, \gamma_b\, D\, K \tan \varphi_s \frac{D}{2}. \tag{5-7}$$

This equation can be solved for depth in terms of the other parameters using the quadratic equation. The result is:

$$D = \frac{-(I + L) \pm [(I + L)^2 + 4J(Q + \text{Bias})]^{1/2}}{2J} \tag{5-8}$$

where $\quad I = A_{ff}\, N_q\, \gamma_b$
$\quad\quad L = A_{fs}\, \gamma_b\, K \tan \varphi_s$
$\quad\quad J = (1/2)a_s\, \gamma_b\, K \tan \varphi_s.$

TABLE 3.   RATIO OF KEYING DISTANCE* TO FLUKE LENGTH

| Type of fluke | Ratio of keying distance to fluke length |
|---|---|
| Expandable (finger-like flukes) | 2–3 |
| Rotating plate fluke | 2–3 |
| Screw-in | 0 |
| Eccentric-keying flat-plate fluke | 1–2 |

*Distance measured vertically from fluke tip.

5.1.3. *Screw or auger penetration.* Penetration of screw or auger types of anchors can be estimated best by reviewing penetrations achieved in various types of soil.

5.1.4. *Penetration reduction due to fluke keying.* The depth of embedment to be used in a holding capacity calculation is not the penetration depth; it is the penetration depth less the distance required to bring the fluke to fluke length for a variety of fluke types. Multiplying these factors by the fluke length will give an estimate of the distance required to key a fluke. These distances are given in Table 3.

## 5.2. *Holding capacity*

The purpose here is to provide methods for estimating the holding capacity of uplift-resisting anchors in seafloor soils. Holding capacity cannot be estimated analytically in rock and coral. In those materials field tests and general experience must be relied upon.

5.2.1. *Basic holding capacity equation.* The maximum uplift forces that can be applied to direct-embedment anchors without causing the anchors to pull out are identified as the anchor holding capacities. Holding capacity is not a property of a particular anchor, but varies considerably with seafloor type, embedment depth, and method of loading.

It is necessary to subdivide the holding capacity problem into categories. The first subdivision is based on general soil type, of which there are two: cohesive and cohesionless. Cohesive soils are fine-grained plastic materials (clays), and cohesionless soil are coarse-grained nonplastic materials (sands). The second subdivision is based on method of loading. For each general soil type three methods of loading will be considered: short-term static, long-term static, and long-term repeated. Short-term static loading describes the situation in which the anchor is loaded rapidly until breakout occurs. Most field tests have been conducted in this manner, and most of the theoretical results are directed toward it. Long-term holding capacities are usually presented as fractions of the immediate capacity. Long-term static holding capacity refers to the situation in which an anchor pulls out after a constant upward force has been applied over a long period of time. This holding capacity would be associated with moored objects such as submerged buoys. Repeated loading involves a line force that varies considerably with time; it can be approximated by a sinusoidally varying force with a certain period and amplitude. Moored surface buoys and ships can provide this type of force application.

The holding capacity problem has been divided into six categories; they are:
  1. Cohesive soil—short-term static loading
  2. Cohesive soil—long-term repeated loading
  3. Cohesive soil—long-term static loading
  4. Cohesionless soil—short-term static loading
  5. Cohesionless soil—long-term repeated loading
  6. Cohesionless soil—long-term static loading.

The commonly used equation for representing the holding capacities of embedment anchors is:

$$F_T = A(c\bar{N}_c + \gamma_b D \bar{N}_q)(0.84 + 0.16 B/L) \qquad (5\text{-}9)$$

where
$\quad A$ = Fluke area (ft²).
$\quad\quad c$ = Soil cohesion (psf), characteristic strength.
$\quad\quad \gamma_b$ = Buoyant unit weight of soil (pcf).
$\quad\quad D$ = Fluke embedment depth (ft).
$\quad \bar{N}_c, \bar{N}_q$ = Holding capacity factors.
$\quad\quad B$ = Fluke diameter or width (ft).
$\quad\quad L$ = Fluke length (ft).

The equation is relatively general and can be applied to almost any form of loading. However, the holding capacity factors and the cohesion may vary with the loading mode, and they have been found to vary with soil type, density, and relative anchor embedment depth, $D/B$ ($B$ is the fluke width). The major problem of estimating holding capacity is then one of estimating $c$, $\bar{N}_c$, and $\bar{N}_q$.

5.2.2. *Holding capacity prediction procedure.* The general procedural framework presented here is shown by the block diagram of Fig. 71; each item of the diagram is discussed

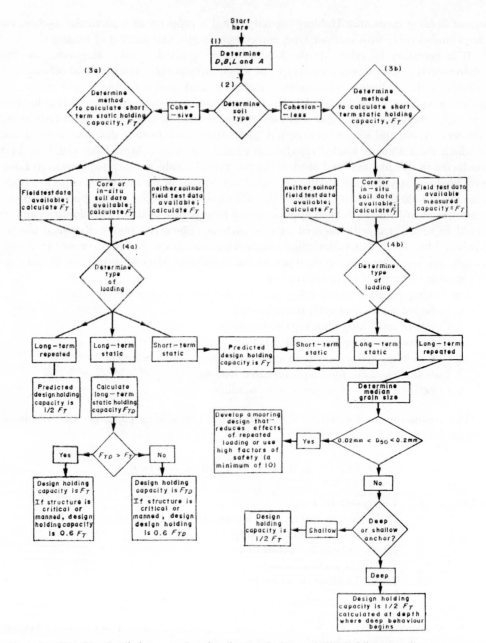

FIG. 71.   Prediction procedure for direct-embedment anchor holding capacity.

briefly below. The numbering system below compares with that of the diagram.

In virtually all cases, an anchor should be installed so as to display "deep" behavior. In all curves of holding capacity or holding-capacity-parameters-vs-depth, there are breaks below which the holding capacity increases less rapidly with increasing depth; this behavior in the lower sections of these plots is termed "deep". It is advantageous to establish

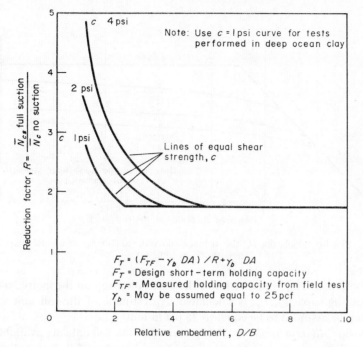

Reduction factor, $R = \dfrac{\overline{N_{cs}} \text{ full suction}}{\overline{N_c} \text{ no suction}}$

c 4 psi

Note: Use $c = 1$ psi curve for tests performed in deep ocean clay

2 psi

c 1 psi

Lines of equal shear strength, c

$F_T = (F_{TF} - \gamma_b\ DA)\ /R + \gamma_b\ DA$
$F_T$ = Design short-term holding capacity
$F_{TF}$ = Measured holding capacity from field test
$\gamma_b$ = May be assumed equal to 25 pcf

Relative embedment, $D/B$

FIG. 72. Reduction factor to be applied to field anchor tests in cohesive soils to account for suction effects.

a "deep" anchor, because errors in locating the anchor, either during installation or because of deformations after installation, do not cause large changes in holding capacity. The anchor is, therefore, more reliable.

A step by step approach for calculating anchor holding capacity is as follows:

(1) *Determine design parameters.* Determine the anchor fluke embedment depth, $D$ (using techniques of Section 5.1), width, $B$, length, $L$, and projected area, $A$.

(2) *Determine soil type.* Determine the general soil type (cohesive or cohesionless). This will be obvious from the visual observation of a bottom sample, even from a very disturbed grab-type sample. In areas far from shore, it may be possible to estimate the type of bottom from a chart of the regional geology. In addition, good geophysical data, if available, may give clues. If at all possible, however, a bottom sample should be obtained.

(3a) *Determine calculation method for cohesive soil.* The short-term static holding capacity for *cohesive soils* can be estimated three different ways depending on the data that are available. One way is based on anchor field test data, the second way on good quality core data in *in situ* strength data, and the third way is for when no soil data or anchor field test data are available.

*Anchor field test data available.* Field tests provide a good means for estimating short-term holding capacities. However, field tests in cohesive soils develop the strength of the soil under the anchor (suction forces) and, therefore, need to be modified to account for these suction forces. If this is not done, unconservative design values will result. Figure 72 can be used to account for the suction effect. Using the relative embedment depth ratio, $D/B$, and an estimate of $c$ (1 psi should be a reasonable value in most cases), a reduction

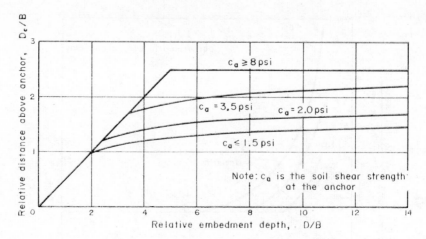

FIG. 73. Plot for calculating $D_c$, the distance above the anchor, at which the characteristic strength, $C_a$, is to be taken.

factor, $R$, is obtained. This is inserted into the equation given on the figure, and the design short-term holding capacity, $F_T$, is calculated. An estimate of the soil unit weight, $\gamma_b$, is needed and can be assumed to be equal to 25 pcf in most cases.

*Core or* in-situ *soil data available.* When core or *in-situ* soil data are available, the short-term static holding capacity can be calculated from Equation 5-9. Some of the values for this equation must be evaluated. Start by making plots of theun drained or vane shear strengthand unit weight distributions. If the strength and density are approximately uniform with depth, then the characteristic strength, $c$, and density, $\gamma_b$, are simply the mean values over the depth range, $D$. If the strength increases approximately linearly with depth from a value of near zero at the seafloor surface, then the plots of Fig. 73 are used to obtain the characteristic strength and density. This is done by first calculating $D/B$ and taking the strength, $c_a$, at depth, $D$ ($D$ is the anchor depth after setting), from the strength profile. Figure 72 is entered with these values, and the quantity $D_c/B$ is determined. $D_c/B$ is the ratio of the distance above the fluke at which the characteristic strength is measured to the fluke width or diameter. The characteristic strength, $c$, and density are then taken as the strength and density a distance $D_c$ above the anchor fluke. For more unusual strength and density profiles, either a conservative uniform or linearly increasing curve should be drawn through the data, or an experienced seafloor soils engineer should be consulted.

Now that $D/B$ and $c$ are known, the parameter $\bar{N}_c$ can be obtained from Fig. 73. $\bar{N}_c$ for cohesive soils is 1. Now that all the values of the parameters have been determined, the short-term holding capacity, $F_T$, can be calculated from Equation 5-9 or from the nomographs, Figs 84, 85 and 86, in Appendix C.

*Neither soil nor anchor field test data available.* When no data are available, soil properties must be assumed to estimate holding capacity. The shear strength and unit weight distributions of Fig. 5.2-5 should be used, and the above steps followed to accomplish this. The procedure can be simplified by using Figs. 78, 79 and 80 in Appendix B where holding-capacities-vs-depth have been plotted for the operative anchors presented in this handbook. If at all possible, however, strengths and densities for the design locations should be measured, and the steps in the above paragraph followed.

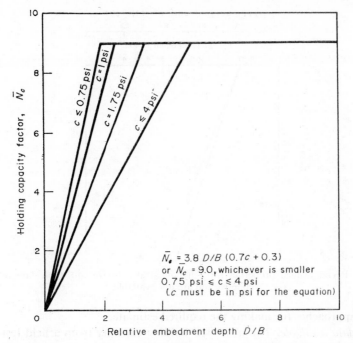

$$\bar{N}_c = 3.8 \, D/B \, (0.7c + 0.3)$$
or $\bar{N}_c = 9.0$, whichever is smaller
$0.75$ psi $\leqslant c \leqslant 4$ psi
($c$ must be in psi for the equation)

Relative embedment depth $D/B$

FIG. 74.   Design curves of holding capacity factor, $N_c$, vs relative embedment depth ($D/B$).

(4a) *Determine type of loading for cohesive soil*. Most anchor trials tests, salvage work, and other projects that require a reaction force for a short period of time are considered to be short-term static loadings. Surface vessels and buoys generally exert a long-term repeated loading condition, although certain designs may convert the repeated load into a virtual long-term static condition. Subsurface buoys, suspended arrays, and other suspended structures exert long-term static loads.

*Short-term static loading*. If the loading is short-term static, the design holding capacity is $F_T$ as determined by the selected method in paragraph 3a above.

*Long-term repeated loading*. If the loading is long-term repeated, the design holding capacity is one half $F_T$ as determined by the selected method of paragraph 3a above. This capacity refers to the characteristic peak repeated load. The rationale for this reduction has been given by Taylor and Lee (1972).

*Long-term static loading*. If the loading is long-term static, the long-term capacity, $F_{TC}$, must be estimated. To do this, parameters for the equation must be evaluated. First, the drained friction angle, $\varphi$, the quantity $D/B$, and the parameter $\bar{N}_q$ are obtained. $\bar{N}_c$ and $c$ are set equal to 0 for long-term conditions. Next, the drained holding capacity, $F_{TD}$, is obtained from Equation 5-9 (substituting $F_{TD}$ for $F_T$). $F_{TD}$ is compared with $F_T$ from paragraph (3a) above, and the lower value is used as a design holding capacity. If the anchored system is critical or manned, the result should be multiplied by 0.6 to account for possible creep effects. This reduction for creep effects has been explained by Taylor and Lee (1972).

(3b) *Determine calculation method for cohesionless soils*. The procedure to be followed in estimating the short-term static holding capacity in *cohesionless soils* depends upon the type of data available. Anchor field test data, core or in-situ soil data, and a lack of data

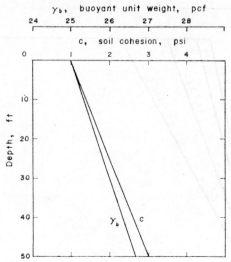

FIG. 75.  Recommended properties for a hypothetical cohesive soil when data on actual cohesive soil are not available.

present three approaches for making the required estimate.

*Field test data available.* The measured holding capacity from a field test can be considered to represent the proper short-term holding capacity, because suction will not be significant in cohesionless soil.

*Core or* in-situ *soil data available.* When core or *in-situ* data are available, Equation 5-9 can be used for estimating the short-term static holding capacity. Values for the parameters in this equation need to be evaluated first. The friction angle, $\varphi$, and the unit weight, $\gamma_b$, in the vicinity of the anchor fluke should be estimated. The parameter $\bar{N}_q$ can be obtained from Fig. 76, given $\varphi$ and $D/B$. $\bar{N}_c$ and $c$ are equal to 0 in a cohesionless soil. The short-term static holding capacity, $F_T$ is now obtained from Equation 5-9 or by using the nomographs, Figs 87, 88 or 89 in Appendix C.

*Soil or field tests data not available.* When no data are available, assume the friction angle to be 30 degrees and the unit weight to be equal to 60 pcf. The procedures of the preceding paragraph can be used with these soil properties to determine $F_T$ by Equation 5-9. The procedure can be simplified by using Figs 81, 82 and 83 in Appendix B where holding-capacities-vs-depth for these soil properties have been plotted for the operative anchors presented in this handbook.

(4b) *Determine type of loading for cohesionless soil.* The type of loading should be determined in a manner identical to that of paragraph (4a).

*Short-term static loading.* If the loading is short-term static, the holding capacity is $F_T$ as determined by the selected method in (4a) above.

*Long-term repeated loading.* If the loading is long-term repeated, the grain size distribution and the relative embedment depth need to be considered. Therefore, a grain size analysis of a soil sample should be performed. If the median grain size ($D_{50}$) is found to lie between 0.02 and 0.2 mm, either a different mooring system design should be developed (i.e. one which reduces effects of repeated loading) or high factors of safety (greater than 10) should be used. For other grain sizes, it is necessary to determine whether the anchor will

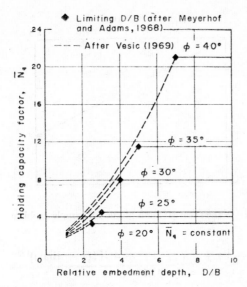

FIG. 76. Holding capacity factor, $\overline{N}_q$, vs relative depth for cohesionless soil, $c = 0$.

be considered "deep" or "shallow". This can be done by referring to Fig. 76 and determining whether the particular range of design parameters places $D/B$ below or above the sharp breaks in the curves. If the anchor is "shallow", the design repeated-load holding capacity is one-half $F_T$ as determined by the selected method in paragraph (4a) above. If the anchor is "deep", it is necessary to calculate the short-term holding capacity at the point where "shallow" behavior changes to "deep". The previous values of $B$, $L$, $\varphi$, and $\gamma_b$ should be used, and paragraph (3b) should be repeated with the new $D/B$. One-half of the short-term holding capacity calculated with these parameters should be used for design purposes.

*Long-term static loading.* When the type of loading is long-term static, the holding capacity is $F_T$ as determined from the method selected under paragraph (3b) above.

### 5.3. Sample problem

A Direct-Embedment Vibratory Anchor (see Section 3.12) with a 3-ft-diameter fluke is to be used in a cohesive soil. The purpose of the anchor is to support a subsurface buoy that is to be in service for several years. A good quality core has been obtained, and the measured vane shear strength profile is given by the curve in Fig. 77. The sensitivity of the soil is 2. The buoyant unit weight was measured and found to be about 35 pcf throughout the profile.

The penetration of the fluke must be determined first, and then the holding capacity can be estimated.

*Penetration.* From Fig. 77 the shear strength or cohesion is shown to increase linearly with depth. Since the buoyant unit weight is constant over the soil profile, the strength can be expressed as a $c/p$ ratio (cohesion to effective overburden pressure). At a depth of 10 ft the cohesion is equal to 2 psi or 288 psf, and the effective overburden pressure is equal to 350 psf (10 feet × 35 pcf). Therefore, the $c/p$ ratio is equal to 0.823. The depth of penetration can be solved with Equation 5-5,

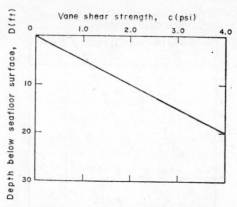

FIG. 77.   Vane shear strength profile for sample problem.

$$D = \frac{-(X + Y) \pm [(X + Y)^2 + 4\,W(Q + \text{Bias})]^{1/2}}{2\,W} \qquad (5\text{-}5)$$

where   $X = A_{fs}\,(c/p)\gamma_b$
   $Y = A_{ff}\,N_c\,(c/p)\gamma_b$
   $W = (1/2)a_s\,(1/S_t)\,(c/p)\gamma_b$
and   $N_c = 9$.
From Beard (1973),
   $A_{fs} = 18.4\ \text{ft}^2$
   $A_{ff} = 0.5\ \text{ft}^2$
   $a_s = 0.813\ \text{ft}^2/\text{ft}$
   $Q = 12{,}500\ \text{lb}$
   Bias $= 540\ \text{lb}$.
and from the soil data,
   $c/p = 0.823$
   $\gamma_b = 35\ \text{pcf}$
   $S_t = 2$.
Therefore,
   $X = 18.4\,(0.823)\,(35) = 530$
   $Y = 0.5\,(9)\,(0.823)\,(35) = 130$
   $W = (1/2)\,0.813\,(1/2)\,(0.823)\,(35) = 5.85$.

$$D = \frac{-(530 + 130) \pm [(530 + 130)^2 + 4(5.85)\,(12{,}500 + 540)]^{1/2}}{2(5.85)}$$
$$= 17.2\ \text{ft and} - 130\ \text{ft}.$$

Since penetrations are positive, the penetration is 17.2 ft. For this type of fluke (an eccentric keying flat plate), the ratio of keying distance to fluke length is taken as 1. The fluke keying distance is then 3 ft (1 times the fluke length). Therefore, the embedment depth to be used in the holding capacity calculations is $17.2 - 3.0 = 14.2$ ft.

   *Holding capacity.* The step identifications by numbers in parentheses are the same as those under Section 5.2.2.

(1) *Determine design parameters.*

$D = 14.2$ ft

$B = 3$ ft

$L = 3$ ft

$A = 6.1$ ft$^2$ (from Beard, 1973).

(2) *Determine soil type.*

The soil is cohesive.

(3a) *Determine calculation method for cohesive soil.* Core or *in situ* soil data available. Therefore, the second method listed under (3a) can be used. First, the characteristic soil strength must be determined. This can be done by using Fig. 73. $D/B$ is calculated to be $14.2/3 = 4.7$, and the strength at the anchor is calculated by multiplying the $c/p$ ratio by the effective overburden pressure at that depth $(\gamma_b D)$, which gives $c_a = 0.825 \times 35 \times 14.2 = 409$ psf. To use Fig. 73, $c$ must be in psi (409 psf $\times$ 1 psi/144 psf = 2.84 psi). From Fig. 73, $D_c/B$ is estimated to be 1.7. Multiplying by $B$, $D_e$ is determined to be 5.1 ft. This is the distance above the anchor at which point the characteristic strength, $c$, is to be determined. At a depth of 9.1 ft (14.2 − 5.1), $c$ is then $c/p \times \gamma_b \times D = 0.823 \times 35 \times 9.1 = 262$ psf or 1.82 psi. Now $\bar{N}_c$ can be determined from Fig. 74 where $D/B = 4.7$ and $c = 1.82$; $\bar{N}_c = \bar{N}_q$ for cohesive soils is 1.

Now the short-term static holding capacity can be calculated from Equation 5-9.

$F_T = A (cN_c + \gamma_b D\bar{N}_q) (0.84 + 0.16 B/L)$

$\quad = 6.1 [(262) (9) + (35) (14.2) (1)] [0.84 + 0.16 (3/3)]$

$\quad = 17,400$ lb.

This is the estimated short-term static holding capacity.

(4a) *Determine type of loading.* The load will be applied for several years from a submerged buoy and can be considered a long-term static load.

Initially it will be a short-term static load. Therefore, the design short-term holding capacity is $F_T$ or 17,400 lb.

For the long-term static loading the following procedure is used. The friction angle, $\varphi$, was not determined by laboratory tests, and, therefore, a conservative value of 25 degrees will be used. Using $D/B = 4.7$ and $\varphi = 25$ degrees, Fig. 76 is used to obtain $\bar{N}_q$ which is equal to 4.5. $\bar{N}_c$ and $c$ are equal to 0 for long-term conditions. Now Equation 5-9 can be used to find the long-term static holding capacity, $F_{TD}$.

$F_{TD} = A (c \bar{N}_c + \gamma_c D \bar{N}_q) (0.84 + 0.16 B/L)$

$\quad = 6.1 [0 + (35) (14.2) (4.5)] [0.84 + 0.16 (3/3)]$

$\quad = 13,600$ lb.

$F_T$ is larger than $F_{TD}$, and, therefore, $F_{TD}$ is the design holding capacity. If the buoy were especially critical, the design holding capacity would be multiplied by 0.6 to account for possible creep effects.

*Answer.* The design holding capacity is 13,600 pounds.

*Acknowledgement*—Grateful appreciation is extended to Mr. J. E. Smith for his detailed review and editing of the handbook and for his assistance in organizing the handbook into a presentable format.

## 6. REFERENCES, BIBLIOGRAPHY, AND PATENTS

6.1. *References*

Beard, R. M. 1973. Direct embedment vibratory anchor, Naval Civil Engineering Laboratory. Technical Report R-791, Port Hueneme, CA, Jun 1973. (AD 766103).

Brackett, R. L. and Parisi, A. M. 1975. Development test and evaluation of a handheld hydraulic rock drill and seafloor fasteners for use by divers, Civil Engineering Laboratory, Technical Report R-, Port Hueneme, CA (to be published).

Brown, G. A. and Nacci, V. A. 1971. "Performance of hydrostatic anchors in granular soils", in Preprints, Third Annual Offshore Technology Conference, Houston, TX, Apr 19–21, 1971. Dallas, TX, Offshore Technology Conference, Vol 2, 1971, pp 533–542. (Paper no. OTC 1472).

Dantz, P. A. 1968. The padlock anchor, a fixed-point anchor system, Naval Civil Engineering Laboratory Technical Report R-577, Port Hueneme, CA, May 1968. (AD 669113).

Dantz, P. A. and Ciani, J. B. 1967. Piston velocities of a single-impulse, deep-ocean hydrostatic ram, Naval Civil Engineering Laboratory, Technical Note N-948, Port Hueneme, CA, Dec 1967. (AD 831131).

Frohlich, H. and McNary, J. F. 1969. "A hydrodynamically actuated deep sea hard rock corer", *Marine Tec. Soc. J.* Vol 3, no. 3, May 1969, pp. 53–60.

Institut Francais du Petrole, des Carburants et Lubricants (1970). Subseas vibro-driver, Catalog Reference 17928A, Rueil Malmaison, France, Mar 1970.

Lair, J. C. 1967. Investigation of embedding an anchor by the pulse-jet principle, Naval Civil Engineering Laboratory, Contract Report CR-68.008. Torrance, CA, Sea-Space Systems, Inc., Oct 1967. (Contract no. NBy-62225).

Meyerhof, G. G. and Adams, J. I. 1968. "The ultimate uplift capacity of foundations", *Can. Geotec. J.* Vol 5, no. 4, Nov 1968, pp. 225–244.

Parisi, A. M. and Brackett, R. L. 1974. Development, test and evaluation of an underwater grout dispensing system for use by divers, Civil Engineering Laboratory, Technical Note N-1347, Port Hueneme, CA, Jul 1974. (AD 786350).

Rossfelder, A. M. and Cheung, M. C. 1973. Implosive anchor feasibility study, Report on Contract no. N62477-73-C-0429, Eco Systems Management Associates, La Jolla, CA, Nov 1973.

Schmid, W. E. 1969. Penetration of objects into the ocean bottom (state of the art), NCEL Contract Report CR 69.030. Princeton, NJ, W. E. Schmid, Mar 1969. (Contract no. N62399-68-C-0044). (AD 695484).

Smith, J. E. 1966. Investigation of embedment anchors for deep ocean use, Naval Civil Engineering Laboratory, Technical Note N-834, Port Hueneme, CA, Jul 1966.

Stevenson, H. J. and Venezia, W. A. 1970. Jetted-in marine anchors, Naval Civil Engineering Laboratory, Technical Note N-1082, Port Hueneme, CA, Feb 1970. (AD 704488).

Taylor, R. J. and Lee, H. J. 1972. Direct embedment anchor holding capacity, Naval Civil Engineering Laboratory, Technical Note N-1245, Port Hueneme, CA, Dec 1972. (AD 754745).

True, D. G. 1975. Penetration into seafloor soils, Civil Engineering Laboratory, Technical Report R-822, Port Hueneme, CA, May 1975.

Wang, M. C., Nacci, V. A. and Demars, K. R. 1974. Vertical breakout behavior of the hydrostatic anchor, CEL Contract Report CR 74.005. Kingston, RI, University of Rhode Island, Feb 1974. (Contract no. N62399-72-C-0005) (AD 775658).

Vesic, Aleksandar S. 1969. "Breakout resistance of objects embedded in ocean bottom", in Civil Eng in the Oceans II, ASCE Conference Miami Beach, FL, Dec 10-12, 1969. New York, ASCE, 1970, pp. 137–165.


6.2. *Bibliography*

A. B. Chance Co. 1954. Retractable-reusable anchor, Rome Air Development Center, Griffis Air Force Base, NY, 1954. (Contract AF-30-602)-388).

A. B. Chance Co. 1969. Encyclopedia of anchoring, Bulletin 424-A, Centralia, MO, 1969.

A. C. Electronics. Defense Research Laboratories. 1969. Technical Manual for Project BOMEX free-fall anchor systems, Manual no. OM69-01. Santa Barbara, CA, Feb 1969.

Adams, J. I. 1963. Uplift tests on model anchors in sand and clay, unpublished report, Research Division, the Hydro-electric Power Commission of Ontario, Canada, Sep 1963. (*Unverified*).

Adams, J. I. and Hayes, D. C. 1967. "Uplift capacity of shallow foundations", *Ontario Hydro Res. Quarterly*, Vol 19, no. 1, 1967, pp. 1–13.

Adams, J. I. and Klym, T. W. 1972. "A study of anchorages for transmission tower foundations", *Can. Geotec. J.* Vol 9, no. 1, Feb 1972, pp. 89–104.

Aerojet-General Corporation. Proposal: Development of three types of propellant activated underwater anchors, Downey, CA, May 1961.

Aircraft Armaments, Inc. (1960). Feasibility investigation of a propellant actuated underwater anchor, Report no. ER-1966, Cockeysville, MD, Mar 1960. (Contract DA-36-034-507-ORD) (AD 234685).

ALI, M. S. 1968. Pullout resistance of anchor plates and anchor piles in soft bentonite clay, Duke University, Soil Mechanics Series no. 17, Durham, NC, 1968, p. 50. (Also MS thesis, Duke University).

American Electric Power Service Corporation, Report of anchor test, 1963. (*Unverified*).

Anchoring and dynamic positioning, *Ocean Industry*, Vol 1, no. 4, Aug 1966, pp. 1A-16A. Special section:

GRAHAM, J. R. *Mooring techniques: a discussion of problems and knowledge concerning station keeping in the open sea*, pp. 1A–5A.

SCHANER, D. S. *Wire line tension measurement: demand grows for devices to monitor loads on ships and semi-submersibles*", pp. 6A–10A.

FOSTER, K. W. *Dynamic anchoring: man's solution to positioning vessels in deeper waters without use of anchors*, pp. 11A–13A.

KORKUT, M. D. and HERBERT, E. J. *Determining the catenary: presenting in condensed, usable form equations to find anchor chain curve*, pp. 14A–16A.

BAKER, W. H. and KONDNER, R. L. (1966). Pullout load capacity of a circular earth anchor buried in sand, *Highway Research Record* 108, 1966, pp. 1–10.

BALLA, A. 1961. The resistance to breaking-out of mushroom foundations for pylons, in *Proceedings of Fifth International Conference of Soil Mechanics Foundation Engineering, Paris*, 17–22 Jul, 1961, Paris, Dunod, 1961, Vol 1, pp. 569–576.

BAYLES, J. J. and JOCHUMS, R. E. (1965). Underwater mooring systems, Naval Civil Engineering Laboratory, Technical Note N-662, Port Hueneme, CA, Apr 1965. (AD 461146).

BEMBEN, S. M. and KALAJIAN, E. H. (1969. "The vertical holding capacity of marine anchors in sand", *Proceedings, Civil Engineering in the Oceans II, ASCE Conference*, Miami Beach, FL, Dec 10–12, 1969. New York, American Society of Civil Engineers, 1970, pp. 117–136.

BEMBEN, S. M., KALAJIAN, E. H. and KUPFERMAN, M. 1973. "The vertical holding capacity of marine anchors in sand clay subjected to static and cyclic loading", in Preprints, 1973 Offshore Technology Conference, Houston, TX, Apr 30–May 2, 1973. Dallas, TX, Offshore Technology Conference, 1973, Vol 2, pp. 871–880. (Paper no. OTC 1912).

BEMBEN, S. M. and KUPFERMAN, M. 1974. The behavior of embedded marine flukes subjected to static and cyclic loading. Report on Contract N62 399-72-C-0018. Amherst, MA, University of Massachusetts, 1974.

BEMBEN, S. M., KUPFERMAN, M. and KALAJIAN, E. H. 1971. The vertical holding capacity of marine anchors in sand and clay subjected to static and cyclic loading, Naval Civil Engineering Laboratory, Contract Report CR 72.007. Amherst, MA, University of Massachusetts, Nov 1971. (Contract no. N62399-70-C-0025) (AD 735950).

BHATNAGAR, R. S. 1969. Pullout resistance of anchors in silty clay, Duke University, Soil Mechanics Series no. 18, Durham, NC, 1969, p. 44. (Also MS thesis, Duke University).

BRADLEY, W. D. 1963. Field tests to determine the holding capacity of explosive embedment anchors, Naval Ordnance Laboratory, Technical Report 63-117, White Oak, MD, Nov 1963. (AD 422984).

BROMS, B. 1965. Design of laterally loaded piles, Proceedings American Society of Civil Engineers, *Journal of the Soil Mechanics and Foundations Engineering Division*, Vol 91, no. SM3, May 1965, pp. 79–99.

BROMS, B. 1964. Lateral resistance of piles in cohesive soils, American Society of Civil Engineers, *Journal Proceedings of the Soil Mechanics and Foundations Engineering Division*, Vol 90, no. SM2, Mar 1964, pp. 27–63.

CAMERON, I. 1969. Offshore mooring devices, *Petroleum Review*, Vol 23, no. 270, Jun 1969, pp. 169–173.

CHRISTIANS, J. A. 1968. Development of multi-leg mooring system, phase D. Designs and layout, Army Mobility Equipment Research and Development Center, Report 1909-D, Fort Belvoir, VA, Apr 1968.

CHRISTIANS, J. A. and MEISBURGER, E. P. 1967. Development of multi-leg mooring system, phase A explosive embedment anchor, Army Mobility Equipment Research and Development Center, Report 1909-A, Fort Belvoir, VA, Dec 1967.

CHRISTIANS, J. A. and MARTIN, E. H. 1967. Development of multi-leg mooring system, phase B explosive embedment anchor fuze, Army Mobility Equipment Research and Development Center, Report 1909-B, Fort Belvoir, VA, Sep 1967. (AD 822168).

Cleveland Pneumatic Industries (1971). Embedment anchor development program. Report 4607-F, Aug 1971. (*Unverified*).

COLP, J. L. and HERBICH, JOHN B. 1972. Effects of inclined and eccentric load application on the breakout resistance of objects embedded in the seafloor, Texas A&M University, Sea Grant Publication no. 72-204, May 1972. (Texas A&M University, Engineering Experimental Station, Rept no. 153-COE).

DANTZ, A. 1966. Light-duty, expandable land anchor (30,000-pound class), Naval Civil Engineering Laboratory, Technical Report R-472, Port Hueneme, CA, Aug 1966. (AD 640232).

DEHART, R. C. and URSELL, C. R. 1967. Force required to extract objects from deep ocean bottom, Report on Contract Nonr-336300, Southwest Research Institute, San Antonio, TX, Sept. 1967, p. 9.

Delco Electronics (1971). A proposal to furnish mooring systems for Project HARPOON, Proposal no. 171-44. Santa Barbara, CA, Aug 1971.

DEMARS, K. R., NACCI, V. A. and WANG, M. C. 1972. Behavior of the hydrostatic anchors in sand, Naval Underwater Systems Center, Report on Contract no. N66604-71-C-0080, Newport, RI, May 1972. (*Unverified*).

Department of the Navy, Bureau of Yards and Docks (1962). Design manual DM-26: Harbor and coastal facilities, Washington, DC, 1962. (Superseded by 1968 ed.).

DOHNER, J. A. 1966. Field tests to determine the holding powers of explosive embedment anchors in sea bottoms, Naval Ordnance Laboratory, Technical Report no. 66-205, White Oak, MD, Oct 1966.

DOWDING, CHARLES. 1970. Anchor-clay-soil interaction, Unpublished term paper, University of Illinois, Urbana, IL, 1970. (*Unverified*).

DRUCKER, M. A. 1934. "Embedment of poles, sheeting, and anchor piles", *Civil Engineering*, Vol 4, no. 12, Dec 1934, pp. 622-626.

ERDEN, S. 1971. A study of the extent of the zone of disturbance of anchors in loose soils, MS thesis, University of Massachusetts, Amherst, MA, May 1971. (*Unverified*).

ERICKSON, F. L. 1972. Explosive embedment anchor development program, Magnavox Company, Report No. FWD71-115, Fort Wayne, IN, Nov 1972. (Contract no. DOT-CG-04468-A).

ESQUIVAL-DIAZ, R. F. 1967. Pullout resistance of deeply buried anchor in sand, Duke University (Soil Mechanics Series no. 8), Durham, NC, 1967. (Also MS thesis, Duke University).

FOTIYERS, N. W. and LITKIN, V. A. "Design of deep anchor plates", Soil Mechanics and Foundation Engineering, no. 5, p. B-10, Plenum, New York. (*Unverified*).

FOX, D. A., PARKER, G. F. and SUTTON, V. J. R. (1970. "Pile driving into North Sea boulder clays", in Preprints, Second Annual Offshore Technology Conference, Houston, TX, Apr 22-24, 1970, Dallas, TX, Offshore Technology Conference, Vol I, 1970, pp. 535-548. (Paper no. OTC 1200).

GOLAIT, A. V. (1967). Model studies on the breaking out resistance of pile foundations with enlarged bases, Unpublished MS thesis, India Institute of Technology, Bombay, 1967. (*Unverified*).

GORDON, D. T. and CHAPLER, R. S. 1972. Vibratory emplacement of small piles, Naval Civil Engineering Laboratory, Technical Note N-1251, Port Hueneme, CA, Dec 1972. (AD 906997).

HALEY and ALDRICH (Consulting Engineers). 1960. Investigations of pull-out resistance of universal ground anchors. Laconia Malleable Iron Co., Laconia, NH, File No. 60-411, 1960. (*Unverified*).

HANNA, T. H. (1968). "Factors affecting the loading behavior of inclined anchors used for the support of tie-back walls," *Ground Engineering*, Vol 1, no. 5, Sept. 1968, pp. 38-41.

Harvey Alum Co. (1966). Test and evaluation of EAW-20 explosive earth anchor system, Marine Corps Landing Force Development Center, Project No. 51-52-01B, Quantico, VA, 1966. (*Unverified*).

HARVEY, R. C. and BURLEY, E. 1973. "Behavior of shallow inclined anchorage in cohesionless sand." *Ground Engineering*, Vol 6, no. 5, Sept. 1973, pp. 48-55.

HEALY, A. 1971. "Pullout resistance of anchors buried in sand," ASCE Proceedings, *J. Soil Mechanics Foundations Dev.* Vol 97, no. SM 11, Nov 1971, pp. 1615-1622.

HOLLANDER, W. L. 1958. "Earth anchors may help you prevent pipe flotation at river crossings or in swamps," *Oil and Gas J.* Vol 56, no. 21, May 26, 1958, pp. 98-101.

HOLLANDER, W. L. and MARTIN, R. 1961. "How much can a guy anchor hold?" *Electric Light and Power*, Vol. 39, no. 5, Mar 1, 1961, pp. 41-43.

HOWAT, M. D. 1965. The behavior of earth anchorages in sand, MS thesis, University of Bristol, 1965. (*Unverified*).

HSIEH, T. Y. and TURPIN, F. J. 1965. Experimental investigation of suction cup anchors, Hydronautics, Inc., Report no. TR-519-1, Sep 1965. (Contract no. Nonr-484500) (AD 803801L).

HUECKEL, S. 1957. Model tests on anchoring capacity of vertical and inclined planes, in *Proceedings of the Fourth International Conference on Soil Mechanics and Foundation Engineering, London, 12-24 Aug 1957*. London, Butterworths Scientific Publications, Vol 2, 1957, pp. 203-206.

JOHNSON, V. E., ETTER, R. J. and TURPIN, F. J. 1967. "Suction cup anchors for underwater mooring and handling," paper presented at American Society of Mechanical Engineers, Petroleum Mechanical Engineering Conference, Philadelphia, PA, Sept. 17-20, 1967.

KALAJIAN, E. H. 1971. The vertical holding capacity of marine anchors in sand subjected to static and cyclic loading, Ph.D. thesis, University of Massachusetts, Amherst, MA, 1971.

KALAJIAN, E. H. and BEMBEN, S. M. 1969. The vertical pullout capacity of marine anchors in sand, Report no. UM-69-5, University of Massachusetts, School of Engineering, Amherst, MA, Apr 1969. (AD 689522).

KANAYAN, A. S. 1963. Analysis of horizontally loaded pipes, *Soil Mechanics and Foundation Engineering*, no. 2, Mar-Apr 1963. (*Unverified*).

KARAFIATH, L. and BEKKER, M. G. 1957. An investigation of gun anchoring spades under the action of impact loads, Army Tank-Automotive Command, Report no. 19, Warren, MI, Oct 1957. (AD 156419).

KENNEDY, J. L. 1969. This lightweight, explosive-set anchor can stand a big pull, *Oil and Gas J*. Vol 67, no. 16, Apr 21, 1969, pp. 84–86.

KHADILKAR, B. S., PARADKAR, A. K. and GOLAIT, Y. S. 1971. Study of rupture surface and ultimate resistance of anchor foundations, in *Proceedings of the Fourth Asian Regional Conference on Soil Mechanics and Foundation Engineering*, Bangkok, 26 Jul–1 Aug 1971. Bangkok, Asian Institute of Technology, 1971, Vol 1, pp. 121–127; discussion, Vol 2, pp. 139–140.

KUPFERMAN, M. 1971. The vertical holding capacity of marine anchors in clay subjected to static and cyclic loading, MS thesis, University of Massachusetts, Amherst, MA, 1971. (*Unverified*).

KWASNIEWSKI, J. and SULIKOWSKA, L. 1964. Model investigations on anchoring capacity of vertical cylindrical plates, in *Proceedings of the Seminar on Soil Mechanics and Foundation Engineering*, Lodz, 1964. (*Unverified*).

LANGLEY, W. S. 1967. Uplift resistance of groups of bulbous piles in clay, MS thesis, Nova Scotia Technical College, Halifax, NS, 1967. (*Unverified*).

LARNACH, W. J. 1972. Pullout resistance on inclined anchors, *Ground Engineering*, Vol 5, no. 4, Jul 1972, pp. 14–17.

LEE, H. J. 1972. Unaided breakout of partially embedded objects from cohesive seafloor soils, Naval Civil Engineering Laboratory, Technical Report R-755, Port Hueneme, CA, Feb 1972. (AD 740751).

LIU, C. L. 1969. Ocean sediment holding strength against breakout of embedded objects, Naval Civil Engineering Laboratory, Technical Report R-635, Port Hueneme, CA, Aug 1969. (AD 692411).

MACDONALD, H. F. 1963. Uplift resistance of caisson piles in sand, MS thesis, Nova Scotia Technical College, Halifax, NS, 1963. (*Unverified*).

Magnavox Co. Explosive embedment penetrometer system, Final Report No. TP 4912. Fort Wayne, IN. (*Unverified*).

Magnavox Co., Government and Industrial Division, Self embedment anchor developments. Urbana IL. (*Unverified*).

MARDESICH, J. A. and HARMONSON, L. R. 1969. Vibratory embedment anchor system, Naval Civil Engineering Laboratory, Contract Report CR-69.009. Long Beach, CA, Ocean Science and Engineering, Inc., Feb 1969. (Contract no. N62399-68-C-0008) (AD 848920L).

MARIUPOLSKII, L. G. 1965. The bearing capacity of anchor foundations, *Soil Mechanics and Foundation Engineering*, Vol 3, no. 1, Jan–Feb 1965, pp. 26–32. (*Unverified*).

MARKOWSKY, M. and ADAMS, J. I. 1961. Transmission towers anchored in muskeg, *Electrical World*, Vol 155, no. 8, Feb 20, 1961, pp. 37–37, 68.

MATLOCK, M. and REESE, L. C. 1962. Generalized solutions for laterally loaded piles, *ASCE Transactions*, Vol 127, pt I, 1962, pp. 1220–1251.

MATSUO, M. 1967. Study on the uplift resistance of footing (I), Soils and Foundations, Japan, Vol 7, no. 4, Dec 1967, pp. 1–37.

MATSUO, M. 1968. Study on the uplift resistance of footing (II), *Soils and Foundations*, Japan, Vol 8, no. 1, Mar 1968, pp. 18–48.

MAYO, H. C. 1972. "Rapid mooring-construction system," *The Military Engineer*, Vol 64, no. 418, Mar–Apr 1972, pp. 110–111.

MAYO, H. C. 1973a. Explosive embedment anchors for ship mooring. Army Mobility Equipment Research and Development Center, Report 2078, Fort Belvoir, VA, Nov 1973.

MAYO, H. C. 1973b. Explosive anchors for ship mooring, *Marine Technology Society J*. Vol 7, no. 6, Sept. 1973, pp. 27–34.

MCKENZIE, R. J. 1971. Uplift testing of prototype transmission tower footings, in Proceedings of First Australian-New Zealand Conference on Geomechanics, Melbourne, Aug 1971, Vol 1, pp. 283–290.

MEYERHOF, G. G. 1973. The uplift capacity of foundations under oblique loads, *Can. Geotec. J*. Vol 10, no. 1, Feb 1973, pp. 64–70.

MIGLIORE, H. J. and LEE H. J. 1971. Seafloor penetration tests: Presentation and analysis, Naval Civil Engineering Laboratory, Technical Note N-1178, Port Hueneme, CA, Aug 1971. (AD 732369).

MUGA, B. J. 1966. Breakout forces, Naval Civil Engineering Laboratory, Technical Note-863, Port Hueneme, CA, Sept. 1966, p. 24.

MUGA, B. J. 1967. Bottom breakout forces, in *Proceedings Civil Engineering in the Oceans, ASCE Conference*, San Francisco, CA, Sept. 6–8, 1967, pp. 569–600. New York, American Society of Civil Engineers, 1968, pp. 569–600.

MUGA, B. J. 1968. Ocean bottom breakout forces, including field test data and the development of an analytical method, Naval Civil Engineering Laboratory, Technical Report R-591, Port Hueneme, CA, Jun 1968, p. 140 (AD 837647).

Neely, W. J. 1972. Sheet pile anchors: Design reviewed the importance of flexibility in the design of sheet pile anchors in sand, *Ground Engineering*, Vol 5, no. 3, May 1972, pp. 14–16.

North American Aviation, Inc. (1965). Hydrostatic embedment anchor tests, Summary Report NA65H-387, Columbus, OH, 1965. (*Unverified*).

Offshore/Sea Development Corp. (1969). Technical proposal: Mud and rock anchor feasibility and design study, 1969.

Radhakrishna, A. S. and Adams, S. I. 1973. Longterm uplift capacity of augered footings in fissured clay *Can. Geotec. J.* Vol 10, no. 4, Nov 1973, pp. 647–652.

Raecke, D. A. and Migliore, H. J. 1971. Seafloor pile foundations: State-of-the-art and deep-ocean emplacement concepts, Naval Civil Engineering Laboratory, Technical Note N-1182, Oct 1971. (AD 889087).

Redick, T. E. 1962. Anchor study, phase 1A, Naval Air Engineering Laboratory, Naval Air Materiel Center, Report NAEL-ENG-6853, Philadelphia, PA, Apr 1962.

Ridgeway, J. J. 1970. Explosive anchors for sea mooring, *Undersea Technology*, Vol 11, no. 12, Dec 1970, pp. 16–17.

Robinson, F. S. and Christians, J. A. 1967. Development of multi-leg mooring system, phase C. Installation, Army Mobility Equipment Research and Development Center. Report 1909-C, Fort Belvoir, VA, Oct 1967.

Schmidt, B. and Kirstensen, J. P. 1964. The pulling resistance of inclined anchor piles in sand, *Danish Geotec. Institute Bull.* no. 18, 1964.

Schuette, H. W. and Sweeney, P. E. 1969. Mud and rock anchor feasibility and design study for multi-leg mooring system; final report on Phase 1, AAI Corp. Report no. ER-5885, Cockeysville, MD, Sep 1969. (Contract no. DAAK02-69-C-0554) (AD 860147L).

Sherwood, W. G. 1967. Developing a free-fall, deep-sea mooring system, in *Transactions of the Second International Buoy Technology Symposium*, Washington, DC, Sept. 18–20, 1967. Washington, DC, Marine Technology Society, 1967, pp. 19–35.

Smith, J. E. 1954. Stake pile development for moorings in sand bottoms, Naval Civil Engineering Laboratory, Technical Note N-205, Port Hueneme, CA, NOV 1954. (AD 81261).

Smith, J. E. 1955. Evaluation of the EZY Pier-Anchor, Naval Civil Engineering Laboratory, Technical Note N-204, Port Hueneme, CA, Feb 1955. (AD 81220L).

Smith, J. E. 1957. Stake pile tests in mud bottom, Naval Civil Engineering Laboratory, Letter Report L-022, Port Hueneme, CA, Sept. 1957.

Smith, J. E. 1963. Umbrella pile-anchors, Naval Civil Engineering Laboratory, Technical Report R-247, Port Hueneme, CA, May 1963, (AD 408404).

Smith, J. E. 1965. Structures in deep ocean engineering manual for underwater construction, chap 7. Buoys and anchoring systems, Naval Civil Engineering Laboratory, Technical Report R-284-7, Port Hueneme, CA, Oct 1965. (AD 473928).

Smith, J. E. 1966a. Investigation of embedment anchors for deep ocean use, paper presented at American Society of Mechanical Engineers, 66-PET-32.

Smith, J. E. 1966b. Investigation of free-fall embedment anchor for deep ocean application, Naval Civil Engineering Laboratory, Technical Note N-805, Port Hueneme, CA, Mar 1966. (AD 808818L).

Smith, J. E. 1971. Explosive anchor for salvage operations; progress and status, Naval Civil Engineering Laboratory, Technical Note N-1186, Port Hueneme, CA, Oct 1971. (AD 735104).

Smith, J. E. 1972. Explosive anchor for salvage operations; progress and status, addendum, TN-1186A, Jan 1972.

Smith, J. E. and Dantz, P. A. 1963. A perspective on anchorages for deep ocean constructions, Naval Civil Engineering Laboratory, Technical Note N-552, Port Hueneme, CA, Dec 1963. (AD 426202).

Smith, J. E., Beard, R. M. and Taylor, R. J. 1970. Specialized anchors for the deep sea; progress summary, Naval Civil Engineering Laboratory, Technical Note N-1133, Port Hueneme, CA, Nov 1970. (AD 716408).

Sowa, V. A. 1970. Pulling capacity of concrete cast *in situ* bored piles, *Can. Geotec. J.* Vol 7, no. 4, Nov 1970, pp. 482–493.

Spence, W. M. Uplift resistance of piles with enlarged base in clay. MS thesis, Nova Scotia Technical College, Halifax, NS. (*Unverified*).

Stevenson, H. J. and Venezia, W. A. 1970. Jetted-in marine anchors, Naval Civil Engineering Laboratory, Technical Note N-1082, Port Hueneme, CA, Feb 1970. (AD 704488).

Sutherland, H. B. 1965. Model studies for shaft raising through cohesionless soils, in *Proceedings of the Sixth International Conference on Soil Mechanics and Foundation Engineering*, Montreal, 8–15 Sept. 1965. Toronto, University of Toronto Press, 1965, Vol 2, pp. 410–413.

Taylor, R. J. and Beard, R. M. 1973. Propellant-actuated deep water anchor; interim report, Naval Civil Engineering Laboratory, Technical Note N-1282, Port Hueneme, CA, Aug 1973. (AD 765570).

Techniques Louis Menard. Publication P/95: Mooring anchors, Longjumeau, France, 1970. (*Unverified*).

THOMASON, R. A. 1964. Propellant-actuated embedment anchor system, Report on Contract no. P.O. 127/34, Downey, CA, Aerojet-General Corp., Jun 1964.

THOMASON, R. A. 1968. Propellant-actuated embedment anchor, Naval Civil Engineering Laboratory, Contract Report CR 69.026. Downey, CA, Aerojet-General Corp., Nov 1968. (Report no. AGC-3324-01(01)FP) (Contract no. N62399-68-C-0002) (AD 850896).

THOMASON, R. A. and WEDAA, H. W. 1961. Special report: The Chuckawalla anchor, Aerojet-General Corp., Report no. 1327-61(02)PB, Downey, CA, Jun 1961.

TIMAR, J. G. and REMBEN, S. M. 1973.The influence of geometry and size on the static vertical pull-out capacity of marine anchors embedded in very loose, saturated sand, University of Massachusetts, School of Engineering. Report no. UN 73-3, Amherst, MA, Mar 1973. (AD 76123).

TOLSON, B. E. 1970. A study of the vertical withdrawal resistance of projectile anchors, MS thesis, Texas A&M University, College Station, TX, May 1970.

TROFIMENKOV, J. G. and MARIUPOLSKII, L. G. 1965. Screw piles for mast and tower foundations, in *Proceedings of the Sixth International Conference on Soil Mechanics and Foundation Engineering*, Montreal, 8–15 Sept. 1965. Toronto, University of Toronto Press, 1965, Vol 2, pp. 328–332.

TRUE, D. G. 1974. Rapid Penetration into Seafloor Soils, in Preprints, Offshore Technology Conference, Houston, TX, May 6–8, 1974. Dallas, TX, Offshore Technology Conference, 1974, Vol 2, pp. 607–618. (Paper no. OTC 2095).

TRUE, D. G., DRELICHARZ, J. A. and SMITH, J. E. Deep water anchor expedient mooring system, Civil Engineering Laboratory, Technical Note N-, Port Hueneme, CA. (To be published).

TURNER, E. A. 1962. Uplift resistance of transmission tower footings, ASCE Proceedings, *Journal of the Power Division*, Vol 88, no. P02, Jul 1962, pp. 17–33.

WILSON, S. D. and HILTS, D. E. 1967. How to determine lateral load capacity of piles, *Wood Preserving*, Jul 1967. (*Unverified*).

WISEMAN, R. J. 1966. Uplift resistance of groups of bulbous piles in sand, MS thesis, Nova Scotia Technical College, Halifax, NS, 1966. (*Unverified*).

YILMAZ, M. 1971. The behavior of groups of anchors in sand, Ph.D. thesis, University of Sheffield, England, 1971. (*Unverified*).


6.3. *Patents*

ANDERSON, M. H., Explosive operated anchor assembly, U.S. Patent No. 3,207,115, filed June 17, 1963.

ANDERSON, V. C. *et al.*, Pressure actuated anchor, U.S. Patent No. 3,311,080.

BAYLES, J. J., Apparatus for mooring instruments at a predetermined depth, U.S. Patent No. 3,471,877.

BAUER, R. F. *et al.*, Anchoring method and apparatus, U.S. Patent No. 2,891,770.

COSTELLO, R. B. *et al.*, Dynamic anchor, U.S. Patent No. 3,187,705.

EDWARDS, T. B., Vibratory sea anchor driver, U.S. Patent No. 3,417,724, filed Sep 27, 1967.

EWING, W. M. *et al.*, Deep-sea anchor, U.S. Patent No. 2,703,544, Mar 8, 1955.

FEILER, A. M., Embedment anchor, U.S. Patent No. 3,032,000, May 1, 1962.

HALBERG, P. V. *et al.*, Mooring apparatus, U.S. Patent No. 3,291,092, Dec 13, 1966.

MOTT, G. E. *et al.*, Deep water anchor, U.S. Patent No. 3,411,473.

MOTT, G. E. *et al.*, Method for installing a deep water anchor, U.S. Patent No. 3,496,900, Feb 24, 1970.

PANNELL, O. R., Explosive embedment rock anchor, U.S. Patent No. 3,431,880, Mar 11, 1969.

PARSON, E. W. *et al.*, Explosive center hole anchor, U.S. Patent No. 3,401,461, Sep 1966.

THOMASON, R. A. *et al.*, Embedment anchor, U.S. Patent No. 3,154,042.

VINCENT, R. P., Explosively driven submarine anchor, U.S. Patent No. 3,525,187, Aug 25, 1970.

*Appendix A*

## SUPPLEMENTARY TABULAR DATA ON SPECIFIC ANCHOR DESIGNS

Table 4 provides a summary of characteristics of uplift-resisting anchors, including such items as operational and performance characteristics and cost. Holding capacity and penetration data are presented in Tables 5 and 6, respectively. The number of data points is inconsistent between Tables 5 and 6 because in some tests penetration depth was not measured while in others holding capacity was not or could not be recorded.

TABLE 6. TEST PENETRATION DATA

| Agency | Anchor | Soft clay (mud) | | Medium to stiff clay | | Sand or sand/gravel | | Coral and rock | |
|---|---|---|---|---|---|---|---|---|---|
| | | No. of tests | Penetration (ft) | No. of tests | Penetration (ft) | No. of tests | Penetration (ft) | No. of tests | Penetration (ft) |
| Magnavox | Model 1000* | — | 16 | — | 10 (est) | — | 9 | — | 1.3 |
| | Model 2000* | — | 16 | — | 10 (est) | — | 9 | — | 1.3 |
| Edo Western | VERTOHOLD 10K | 1 | 10 | — | — | 14 | 9 to 17 | 21 (coral) | 3 to 7 |
| Teledyne Movible Offshore, Inc. | SEASTAPLE mark 5 | 2 | 24 to 25 | 8 | 9 to 14 | 9 | 2 to 13 | 6 (coral) | 1 to 5 |
| | SEASTAPLE mark 50 | 2 | 40 to 45 | — | — | 1 | 30 | 1 (rock) | 5 |
| U.S. Navy (CEL) | 20K Propellant anchor 100K | 4 | 34 to 47 | — | — | 5 | 9-1/2† to 30 | 4 (rock) | 2 to 3 |
| | Propellant anchor | 5‡ | 34 to 54 | — | — | 3‡ | 5 to 18 | 3 (rock) | 4 to 5 |
| U.S. Army (MERDC) | Model XM-50 | 1 | 41 | 1 | 40 | 5 | 16 to 26 | 2 (coral) | 18 to 23 |
| | Model XM-200 | 7 | 14 to 42 | 19 | 16 to 49 | 8 | 19 to 30 | 5 (coral) | 10 to 21 |
| U.S. Navy (CEL) | Vibratory anchor | 2 | 7 to 15 | 7 | 4 to 16 | 8 | 2 to 16 | — | — |
| Ocean Science and Engineering, Inc. | Model 1 | — | — | 3 | 6 to 11 | — | — | — | — |
| | Model 2000 | — | 40 | — | 40 | — | 40 | — | — |
| Anchoring, Inc. | Model 2-6in. | 1 | 5 | 1 | 10 | — | — | — | — |
| | Model 3-4in. | — | — | 1 | 10 | — | — | — | — |
| | Model 3-6in. | 3 | 10 | — | — | — | — | — | — |
| U.S. Navy (NAVFAC) | Stake pile | | | | | | | | |
| | Class C 8-in. | 2 | 35 | — | — | 6 | 34 to 44 | — | — |
| | Class B 12-in. | 2 | 35 | — | — | 1 | 35 | — | — |
| | Class A 16-in. | 2 | 35 | — | — | — | — | — | — |
| | Umbrella pile | | | | | | | | |
| | Mark III | — | — | — | — | 1 | 19 | — | — |
| | Mark IV | 2 | 13 | — | — | 2 | 17 to 18 | — | — |
| U.S. Navy (CB) | Jetted anchor | — | — | — | — | 22 | 6 to 9 | — | — |

*Magnavox Co. has done extensive testing in simulated laboratory conditions and in on-site situations. The exact number of tests is not known. The figures listed are approximations based on graphs and other data provided by the company.

†This low penetration resulted from use of a reduced propellant charge; penetration in excess of 15 ft would be typical.

‡Original umbrella flukes used.

*Appendix B*

## CURVES FOR SHORT-TERM STATIC HOLDING CAPACITY VS DEPTH

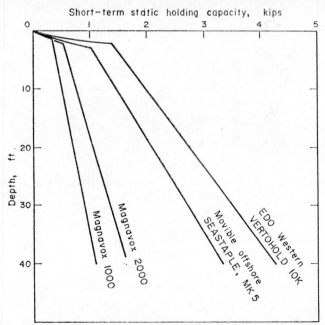

FIG. 78.   Short-term static holding capacity vs depth for small uplift-resisting anchors embedded in the cohesive soil described by Fig. 75.

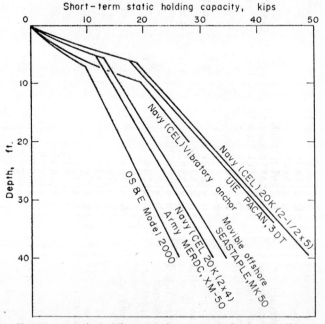

FIG. 79.   Short-term static holding capacity vs depth for intermediate uplift-resisting anchors embedded in the cohesive soil described by Fig. 75.

This appendix presents curves of short-term static holding capacity vs depth for the operative anchors of Section 3 when soil properties must be assumed. Figures 78, 79 and 80 show short-term static holding capacity vs depth for small, intermediate, and large anchors, respectively, when they are to be used in the

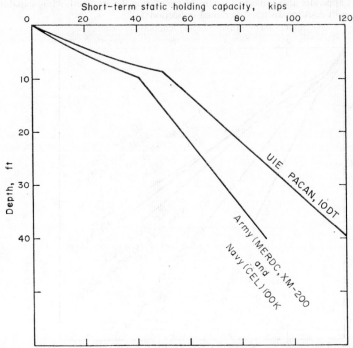

FIG. 80. Short-term static holding capacity vs depth for large uplift-resisting anchors embedded in the cohesive soil described by Fig. 75.

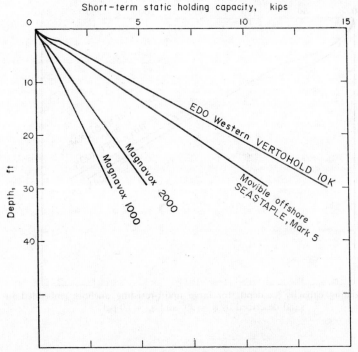

FIG. 81. Holding capacity vs depth for small uplift-resisting anchors embedded in the sand described by $\varphi = 30°$ and $\gamma_b = 60$ pcf.

cohesive soil of Fig. 75. Figures 81, 82 and 83 show short-term static holding capacity vs depth for small, intermediate, and large anchors, respectively, when they are to be used in cohesionless soil where θ = 30 degrees and γb = 60 pcf.

The curves of this appendix also provide a means of comparing the relative holding capabilities of the variety of operative uplift-resisting anchors presented in Section 3.

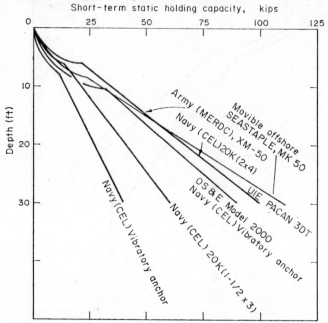

Fig. 82.   Holding capacity vs depth for intermediate uplift-resisting anchors embedded in the sand described by φ = 30° and $\gamma_b$ = 60 pcf.

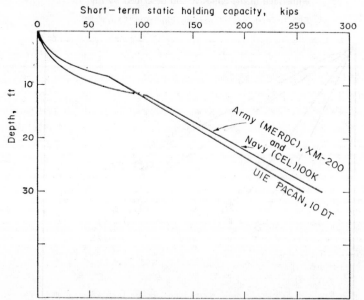

Fig. 83.   Holding capacity vs depth for large uplift-resisting anchors embedded in the sand described by φ = 30° and $\gamma_b$ = 60 pcf.

*Appendix C*

# NOMOGRAPHS FOR CALCULATING HOLDING CAPACITY

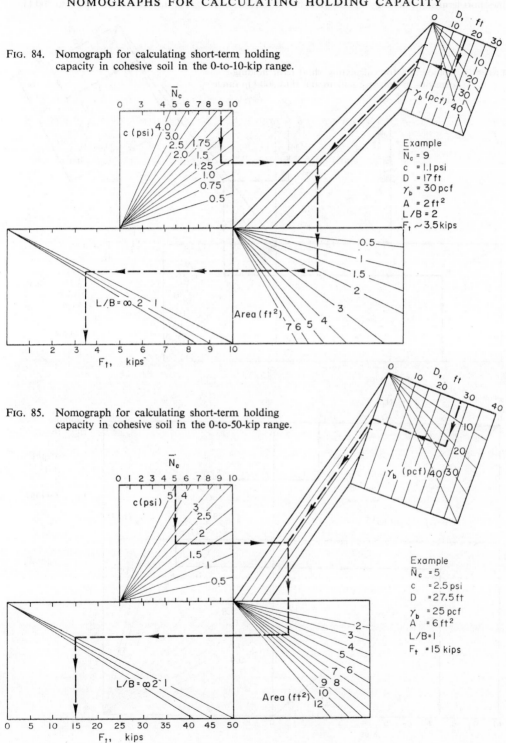

FIG. 84. Nomograph for calculating short-term holding
capacity in cohesive soil in the 0-to-10-kip range.

Example
$\overline{N}_c = 9$
c   = 1.1 psi
D   = 17 ft
$\gamma_b$ = 30 pcf
A   = 2 ft$^2$
L/B = 2
$F_t \sim 3.5$ kips

FIG. 85. Nomograph for calculating short-term holding
capacity in cohesive soil in the 0-to-50-kip range.

Example
$\overline{N}_c$ = 5
c   = 2.5 psi
D   = 27.5 ft
$\gamma_b$ = 25 pcf
A   = 6 ft$^2$
L/B = 1
$F_t$ = 15 kips

The nomographs provide an expedient method for solving the basic holding capacity Equation 5-9 in Section 5 after the parameters in the equation have been evaluated. Figures 84, 85 and 86 are for calculating the short-term static holding capacity in cohesive soils in the ranges of 0 to 10, 0 to 50, and 0 to 200 kips,

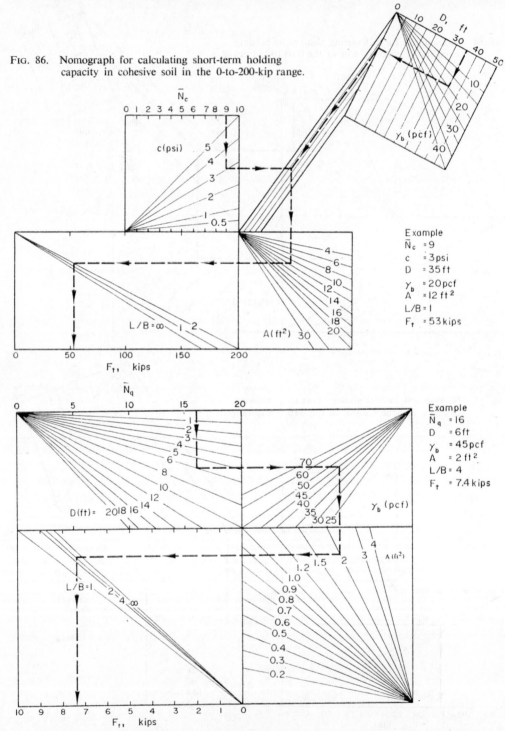

FIG. 86.    Nomograph for calculating short-term holding capacity in cohesive soil in the 0-to-200-kip range.

FIG. 87.    Nomograph for calculating holding capacity in sand in the 0-to-10-kip range.

respectively. Figures 87, 88 and 89 are for calculating the short-term static holding capacity in cohesionless soils in the ranges of 0 to 10, 0 to 100, and 100 to 300 kips, respectively. A sample problem is presented with each nomograph to illustrate usage of the nomograph.

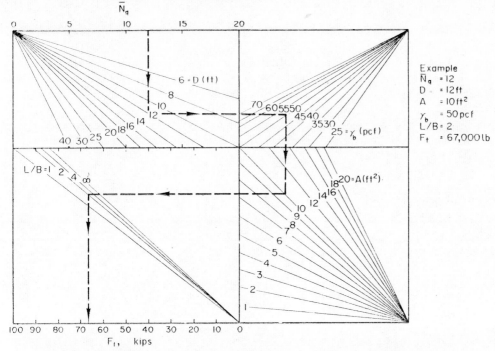

FIG. 88.   Nomograph for calculating holding capacity in sand in the 0-to-100-kip range.

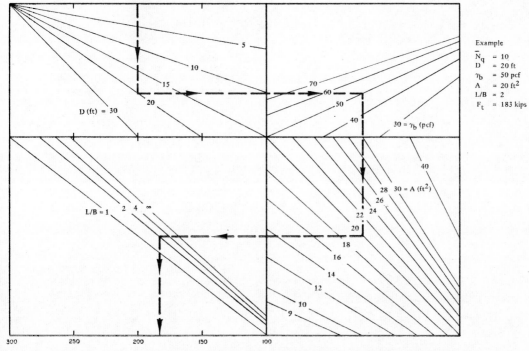

FIG. 89.   Nomograph for calculating holding capacity in sand in the 0-to-300-kip range.

*Ocean Engng.* Vol. 6, pp. 139-167. Pergamon Press 1979. Printed in Great Britain

# PRELIMINARY SELECTION OF ANCHOR SYSTEMS FOR OCEAN THERMAL ENERGY CONVERSION

J. M. ATTURIO, P. J. VALENT and R. J. TAYLOR

Civil Engineering Laboratory, Naval Construction Battalion Center, Port Huenemene, California 93043, U.S.A.

## NOMENCLATURE

| | |
|---|---|
| $B$ | Anchor width, m |
| $\bar{c}$ | Cohesion intercept, Pa |
| $D$ | Pile diameter, m |
| $d$ | Embedment depth, m |
| $H$ | Height of skirt, m |
| $i_q, i_\gamma$ | Bearing capacity factors |
| $P$ | Total force applied at the anchor, N |
| $P_H$ | Horizontal force applied at the anchor, N |
| $P_V$ | Vertical force applied at the anchor, N |
| $Q$ | Bearing capacity, N |
| $Q_e$ | Force required to embed cutting edges, N |
| $R$ | Force or capacity, N |
| $R_A$ | Pullout capacity, N |
| $R_c$ | Submerged weight of square concrete block with height, $H = 0.1B$, N |
| $R_b$ | Lateral resistance of base, N |
| $R_L$ | Total lateral capacity, N |
| $R_p$ | Lateral resistance of passive wedge, N |
| $R_R$ | Required submerged weight, N |
| $R_V$ | Resultant of vertical forces at the base of the cutting edges, N |
| $S$ | Shearing strength, Pa |
| $S_t$ | Soil sensitivity (dimensionless) |
| $W$ | Submerged weight, N |
| $Z$ | Embedded length, m |
| $\beta$ | Mooring line angle with the seafloor, rad. |
| $\bar{\varphi}$ | Effective angle of internal friction, rad. |

## INTRODUCTION

*Purpose*

This report summarizes work accomplished to date on a project entitled, "Anchor Systems for Proposed Ocean Thermal Energy Conversion Power Plants (OTEC)." The Civil Engineering Laboratory (CEL) was to provide performance characteristics of existing anchors, estimated performance characteristics of enlarged versions of these anchors, and predicted performance characteristics of new anchors devised under this study. The work is sponsored by the Division of Solar Energy of the U.S. Energy Research and Development Administration (ERDA).

The first progress report (Valent *et al.*, 1976a) defines probable OTEC anchor environments; specifies upper limits of required holding capacity; and groups all possible seafloor sites into five categories based on soil characteristics.

The second report (Valent *et al.*, 1976b) presents performance characteristics of several possible OTEC anchors.

This report summarizes the findings of the first two publications and also attempts to interpret and use the data in those reports to establish the relative merit of possible OTEC anchors. Finally, this report indicates the next steps to be taken in design of the OTEC anchor.

*Background*

Station-keeping for the OTEC power plant could generally be accomplished by dynamic positioning or by anchoring to the ocean bottom. Several contractors have assessed the feasibility of dynamic positioning using water discharge from the heat exchangers. Such a system could just maintain station in a surface current of 2 km/hr (1 knot) and an 80 km/hr (50 knots) wind (Douglass, 1975). Currents of about 4 km/hr (2 knots) would produce a drift of about 22 km (12 nmi) per day. Dynamic positioning in the Gulf Stream, with its surface current of 10 km/hr (5.5 knots), is probably not feasible because of the large current forces and close proximity of shallow water.

The second station-keeping alternative is to moor the OTEC plant—the method discussed in this article. The mooring could be a single-point or a multipoint configuration (Figs 1 and 2, respectively). The mooring configuration directly affects anchor requirements. For example, increasing the mooring line angle with the seafloor will:

1. Decrease the line length needed
2. Increase tension in the mooring line, uplift at the anchor, and downward force on the OTEC plant
3. Increase the magnitude of dynamic loads at the anchor.

A second consideration about moorings is that increasing the number of mooring legs will not substantially reduce the required strength of the mooring leg. Little *et al.* (1976), indicates that use of a 12-point moor vs a single-point moor reduces the strength required in the individual mooring leg by 40% in 600 m water depth and 55% in 1,500 m water depth (assuming the mooring legs are chain). Thus, in effect, only 2 of the 12 legs are active in resisting the lateral forces applied at any one time. The lines and anchors of each leg of a 12-point moor need be only half as strong as that of a single-point moor in order

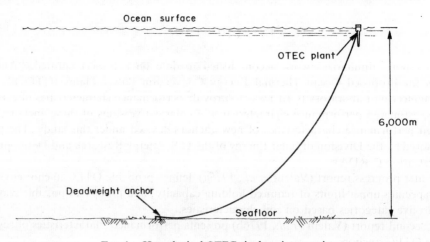

FIG. 1.   Hypothetical OTEC single-point mooring.

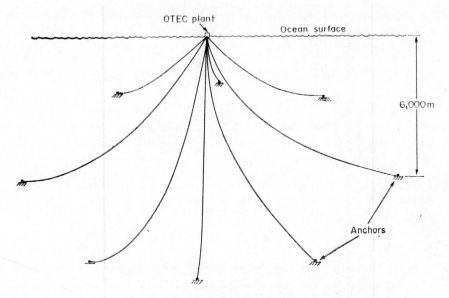

OTEC plant

Ocean surface

6,000 m

Anchors

FIG. 2.   Hypothetical OTEC eight-point mooring.

to maintain the plant on station; however, 12 times the number of legs and approximately six times the weight of line and anchors are required. Thus, from a simple "quantity-of-material" standpoint, the multipoint moor does not appear cost-effective for use with OTEC.

## ENVIRONMENT

*Site categories*

The sediment strength profiles described in Valent *et al.* (1976a) were grouped into three site categories. One shear strength profile was developed to represent each category (Fig. 3). These three site categories, plus two additional ones (one for sands and one for outcropping and near-surface seafloor rock), were used as representative OTEC seafloor environments. The selection, sizing, and comparison of anchor types and combinations refer back to these base-line categories.

Table 1 lists the sediment characteristics described by each category. Each category has been assigned a descriptive word title in Table 1; for example, the sediments of category A can be descriptively referred to as clays and silty clays.

The sands of category D are assumed to be a predominately quartz grain material with a friction angle of 0.5 radian (30°) and a submerged unit weight of 6.8 kN/m³ (43 pcf). Such material under repeated loading may be susceptible to liquefaction. If the median grain size ($D_{50}$) is found to lie between 0.02 and 0.2 mm, then the mooring system should be designed to reduce the effects of repeated loading, or high factors of safety (greater than 10) should be used in the soils design of the anchor.

The material characteristics of site category E—exposed or shallowly buried rock—have not been resolved. The seafloor rock category will be filled in as the need for data develops.

TABLE 1. PROFILE GROUPINGS AND ASSUMED SITE CATEGORY CHARACTERISTICS

| Category | Generalized description | Typical materials | Shear strength, kPa (psi) | Effective friction angle, rad (deg) | Sensitivity | Submerged weight kN/m³ (lb/ft³) | Comments |
|---|---|---|---|---|---|---|---|
| A | Clay, silty, clay* | Pelagic clay, abyssal hill province | 0 | 0.52 (30) | 2-7 | 4.1 (26) | <25% forams |
| | | clay and nannofossil calcareous ooze | 3.45 (0.5) | 0.61 (35) | 5-7† | | |
| | | soft basin material | 0 | 0.66 (37) | 5 | | |
| | | distal turbidite (weak) | 1.38 (0.2) | 0.66 (37) | 2-4 | | |
| | | | 2.76 (0.4) | 0.5 (30) | 3-5 | | |
| B | Clayey silt* | Loose foram ooze, calcareous | 0 | 0.52 (30) | 3-10 | 5.8 (37) | >60% forams |
| | | | 0 | 0.58 (33) | 10 | | |
| | | distal turbidite (strong) | 2.76 (0.4) | 0.52 (30) | 3-5 | | |
| C | Clayey silt, silt* | Proximal turbidite (silt) | 0 | 0.61 (35) | 2-7 | 6.8 (43) | >60% forams |
| | | | 1.71 (0.25) | 0.68 (39) | 2-3 | | |
| | | dense foram ooze, calcareous | 0 | 0.58 (33) | 5-7 | | |
| D | Sand* | Proximal turbidite (sand) | 0 | 0.52 (30) | 10 | 6.8 (43) potentially liquefiable | |
| E | Rock | Seafloor rock, exposed or shallowly buried* | — | — | — | — | — |

*Values are based on engineering judgments.
†For upper 4.6 m; 3 to 5 for below 4.6 m.

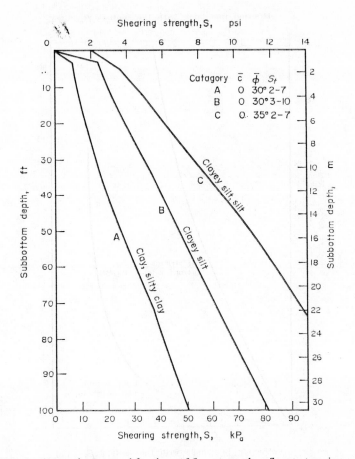

Fig. 3. Properties assumed for three of five assumed seafloor categories.

*Loading*

Anchor concepts for the relatively benign deep ocean environment and for the high energy Gulf Stream type of environment are discussed separately in this report. The assumed maximum horizontal loading for the deep ocean is 18 MN ($4 \times 10^6$ lb) and for the Gulf Stream 180 MN ($40 \times 10^6$ lb). Vertical load at the seafloor depends on the assumed mooring line angle with the seafloor. Figure 4 shows how mooring forces increase as the mooring line angle increases.

In the deep ocean, an assumed maximum mooring line angle of 1.4 radians (80°) results in a vertical force of 102 MN ($22.7 \times 10^6$ lb) at the anchor. The assumed maximum mooring line angle in the Gulf Stream was 0.79 radian (45°), which results in a 180 MN ($40 \times 10^6$ lb) vertical force at the anchor. Figure 5 is a plot of the entire loading envelope, including these maximum values. Only static loads caused by wind and current forces were considered. The dynamic loads from wave action were assumed to be completely damped by the mooring line configuration. Increasing the line angle, thus stiffening the mooring, decreases the validity of this assumption.

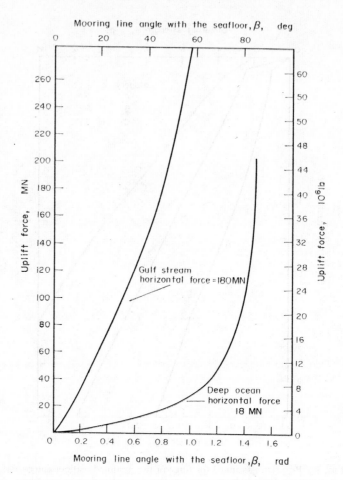

FIG. 4.   Mooring force vs mooring line angle with the seafloor.

## COMPARISONS OF ANCHOR TYPES FOR THE DEEP OCEAN ENVIRONMENT

*Assumed environmental characteristics*

Table 2 describes one of the more demanding deep ocean anchoring sites:

1. The sediments are pelagic clays or distal turbidites, represented by sediment category A.

2. The water depth may reach 6,000 m.

3. A horizontal load of 18 MN (4 × 10⁶ lb) and a vertical load of 0 to 102 MN (22.7 × 10⁶ lb) (depending on mooring line angle at the seafloor) must be resisted by the anchor.

*Comparison criteria*

Several possible deep water OTEC anchors are compared in Table 3, which gives a quick overview of the physical dimensions, advantages, and disadvantages of these anchors. A brief explanation of the comparison criteria used in Table 3 follows.

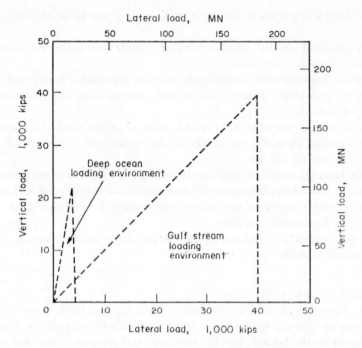

FIG. 5.    Assumed envelope of mooring line forces at the anchor (Jan. 1976).

TABLE 2.    LOADING AND SITE CHARACTERISTICS OF THE DEEP OCEAN ENVIRONMENT

| Characteristics | Description |
|---|---|
| Sea condition | Benign |
| Depth | 6,000 m |
| Bottom soil | Category A: clay, silty clay |
| Assumed upper limits* | |
| Horizontal force | 18 MN (4 × 10⁶ lb) |
| Range of mooring line angles considered | 0.0 to 1.4 rad (0 — 8–80°) |
| Vertical force | 0.0 MN to 101 MN (0.0 to 22.7 × 10⁶ lb) |
| Load direction | Omnidirectional |

*Based on estimates of horizontal loading: 5.6 MN (1.27 × 10⁶ lb) and 8.9 MN (2.0 × 10⁶ lb).

*Description.* The table gives the size of each component of the specified anchor. The mooring line angle for which a particular anchor was designed is also given. For several anchors (e.g. piles) no alteration of anchor size or weight was required for higher line angles. A blank in the table indicates that the characteristic or dimension noted is unchanged. Safety factors, other than those provided by the assumed load, were not included in the designs.

*Performance.* Several qualitative adjectives were used to describe relative anchor performance. Reliability was taken as the probability that an anchor would not fail catastrophically (i.e. allowing the plant to drift freely). It should be noted that the consequence of OTEC breaking away from its mooring, in fact, may not be severe enough to justify overdesign.

The directional load column indicates whether a single anchor may be loaded from only one direction or from any direction.

The repetitive load column qualitatively assesses the anchor's response to a slowly applied, repeated, quasi-static loading such as that resulting from changing current or wind magnitudes and direction.

Efficiency, as used in the table, refers to the ratio of anchor holding capacity to anchor weight in air. Holding capacity was defined as the vector sum of horizontal and vertical mooring line force on the anchor.

*Installation.* Installation was divided into two categories: hardware and procedure. Hardware indicates the special equipment needed to transport or emplace the anchor. Procedure refers to the relative number of operations which must be performed for installation, assuming that the hardware is available.

Cost was not explicitly considered in the comparison, although several of the criteria are directly related to anchor cost.

### Deadweight anchors with cutting edges

*Summary of findings.* A deadweight anchor is basically a large mass which develops lateral resistance by mobilizing soil shear strength. A deadweight anchor with a grid-like skirt arrangement on the bottom uses the increased soil resistance available at deeper soil depths. In effect, the plane of failure (sliding) is driven deeper into the stronger soil, beneath the cutting edge, or skirt, tips. A typical deadweight anchor with keying skirts is shown in Fig. 6. On a deep ocean clay seafloor, a deadweight anchor with skirts may resist up to three times the lateral load of a deadweight anchor without skirts.

Analysis of deadweights in Valent *et al.* (1976b) showed that the optimum value for skirt length in deep ocean clay is 0.1 times the deadweight width. To minimize the possibility of overturning, the maximum deadweight height considered was 0.1 times the width.

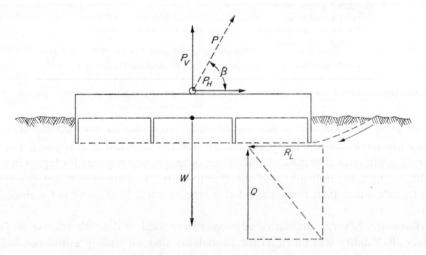

FIG. 6. Deadweight anchor with cutting edge and nonzero line angle, $\beta$.

The lateral load resistance developed by a deadweight with skirts in clay is a function of anchor width ($R_L$ curve in Fig. 7). This total lateral resistance is the sum of the base

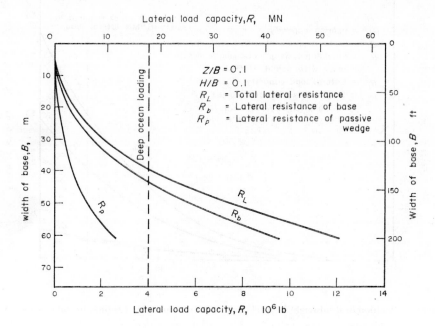

FIG. 7.   Lateral capacity of deadweight anchors vs width in soil category **A**.

resistance ($R_b$) and the passive wedge resistance ($R_p$). Typically, the base resistance is four to five times larger than the resistance in front of the deadweight skirt (passive wedge). Figure 7 indicates that a 40 × 40 m deadweight is required to resist the assumed deep ocean lateral loading of 18 MN (4 × 10$^6$ lb). Once skirts are embedded, the weight on the seafloor serves to resist overturning and uplift forces, not lateral load. For the special case of deadweight height plus skirt length equal to 0.2 times anchor width, the minimum required weight on the seafloor may be specified as follows:

$$R_R = \text{the greater of } (R_L + R_L \tan \beta) \text{ or } Q_e$$

where $R_L$ = total lateral resistance of the soil, N

     $\beta$ = mooring line angle with the seafloor, rad

     $Q_e$ = force required to embed keying skirt, N.

Values for $R_L$ are obtained from Figs 7 or 8. The maximum weight that may be placed on the seafloor is limited by the soil-bearing capacity ($Q$), shown in Fig. 8.

*Relative merits of the deadweight.* The relative merits of a deadweight anchor with cutting edges are compared to various other possible OTEC anchors and summarized in Table 3. Simplicity, reliability, and holding capacity make the deadweight anchor with keying skirts the first choice for OTEC in the deep ocean environment.

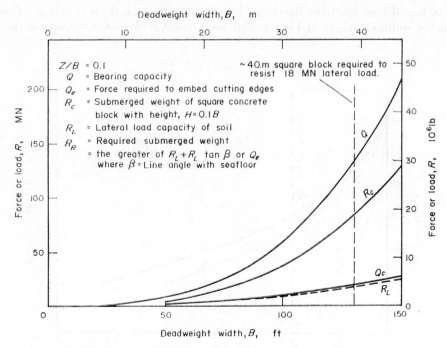

FIG 8.   Relation of deadweight bearing capacity to various other loadings in soil category A.

The deadweights shown in Table 3 were designed from the data presented in the Summary of Findings. Each deadweight is assumed to be fitted with steel cutting edges with heights one-tenth the anchor width, although this is not explicitly noted in Table 3.

The deadweight was judged to be the most reliable of the anchor types listed. Failure would not be catastrophic; rather, it would probably mean a tilting or, at worst, a slow dragging of the anchor.

The deadweight, will be affected least of all the anchors considered by a repetitive loading situation.

Deadweight efficiency remains relatively constant at about 0.5. As noted in Table 3, line angles up to approximately 0.13 rad (8°) can be handled without increasing deadweight mass. Further increases in line angle require additional weight on the seafloor, reducing efficiency.

Installation of a large deadweight system presents some special problems. There are two critical factors: (1) handling components of the dimensions specified and (2) properly applying enough weight to evenly and uniformly embed the skirts. Several methods have been suggested for accomplishing these tasks. One possibility is a heavy lift system capable of carrying the required mass to the site and then reaching down to place the mass on the seafloor, (the Hughes Glomar Explorer has such a capability). Alternatively, the mass could be floated to the site using either internal or external buoyancy tanks. On site, the mass could be lowered by controlled flooding of buoyancy tanks.

Several alternatives exist within each of these plans. For example, a simple box-like frame could be towed to the site, flooded, and placed on the bottom. The weight required to embed the skirts could then be added incrementally or possibly by pouring concrete underwater to fill the box. The authors note that the ability to perform tasks like these in water depths to 6,000 m has not been demonstrated. However, the capability to perform such tasks appears to be a reasonable and relatively logical extension of present near-shore capabilities.

The predicted performance of the basic hypothetical deadweights discussed here indicates that the deadweight anchor with cutting edges is the logical anchor choice for OTEC in the deep ocean.

*Pile anchors*

*Summary of findings*. A simplified single pile type of anchor is shown in Fig. 9.

1. *Axial capacity* (*pullout*). A vertical anchor pile develops its vertical or axial capacity by mobilizing the shearing strength of the adjacent soil. Full mobilization of the soil shearing strength is difficult to achieve in the sensitive calcareous soils known to exist over most of the potential OTEC siting area. In these calcareous soils, piles will have to be grouted

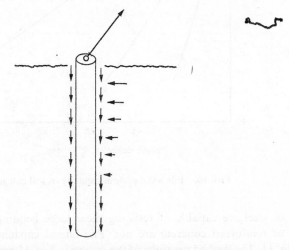

FIG. 9.   Simple pile anchor (develops capacity through skin friction and soil lateral resistance).

in place to develop maximum resistance because capacities of driven piles in this very sensitive soil are notoriously low. The grout annulus injected around the pile would strengthen the pile-to-soil bond as well as increase the strength of the surrounding soil. Pullout resistance depends on the pile surface area, which, of course, varies with pile diameter and length. Pile pullout capacities are plotted against pile length and diameter for category A soil in Fig. 10. A 4.9-m-dia steel pile almost 110 m in length would be sufficient to resist the assumed 102 MN (22.7 × 10⁶ lb) maximum pullout force in the deep ocean, assuming a 1.4 rad (80°) line angle.

2. *Lateral capacity*. When deflected by a lateral load, a pile bears against adjacent soil. The resistance developed is determed by a complex interaction of pile strength and soil strength. Lateral loads are the critical loads in OTEC pile anchor design. The soft ocean bottom sediment allows large lateral deflection of the pile head. This deflection results in a large bending moment which dictates the pile cross section necessary to prevent failure.

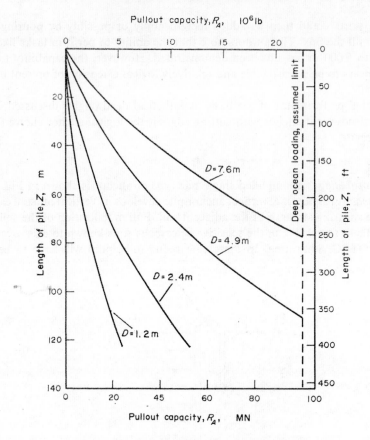

FIG. 10.   Pile axial pullout capacity in soil category A.

Piles made of steel are capable of resisting these large bending moments; piles made of prestressed or reinforced concrete are not. Pile lateral capacity was plotted against pile length in Fig. 11. The dashed portions of the curves in Fig. 11 represent bending failure of a hypothetical pile section. The solid lines represent maximum soil resistance available. Figure 11 shows that the single 4.9-m-dia steel pile (63.5-mm wall thickness) selected above, regardless of length, cannot resist the 18-MN lateral load.

At pile lengths longer than approximately 10 times the pile diameter, bending failure of the pile governs design. Further increases in pile length increase capacity by small amounts. Increasing pile wall thickness does little to enhance bending resistance and leads to uneconomical pile sections. The only practical way to increase lateral capacity is to increase pile diameter or to increase the number of piles per anchor.

Axial and lateral capacity of the typical 4.9-m-dia pile are plotted against pile length in Fig. 12. For a given length, the axial capacity of this pile exceeds its lateral capacity. This demonstrates that commonly used pile sections are not well-suited for resisting high lateral loads. It also shows why pile efficiency (as defined here) increases as the mooring line angle is increased (see Table 3). A pile designed to resist a given lateral load will have sufficient axial capacity to resist pullout forces at high line angles.

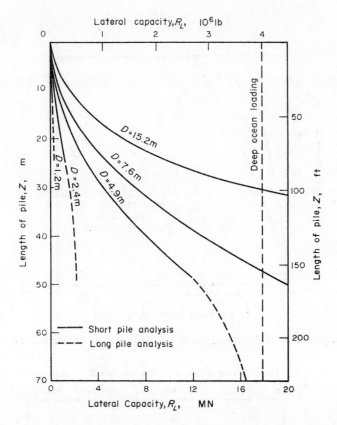

FIG. 11.   Pile lateral capacity in soil category A with head of pile free to rotate.

3. *Repetitive load.* The critical nature of pile lateral capacity is made even worse by repetitive loading. Repetitive, quasi-static loading caused by current and wind variations increases pile deflection and results in a larger bending moment. Thus, the capacity plotted in Fig. 11 must be reduced even further when repetitive loading is considered. Based on model work reported in the literature (Matlock, 1970), the lateral capacity under repetitive loading should be assumed to be approximately 60% of the static load capacity plotted in Fig. 11.

*Relative merits of pile anchors.* The pile system described in Table 3 was designed from the information presented above. Since a single pile of 4.9 m in diameter could not produce the required lateral capacity, a group of three piles connected by a circular pile cap 18.6 m in diameter and 1.0 m thick was specified. The piles were assumed to be grouted in place. As the mooring line angle is increased, pullout force on the pile group increases; but the group is able to withstand considerable pullout force without additional pile length. Therefore efficiency, as calculated herein, increases. The maximum efficiency of 2.7 occurs at 1.3 rad (75°).

Figure 13 compares the efficiency of the pile group to the 40-m square deadweight as the mooring line angle is increased (note that piles are much more efficient at higher line angles). This high efficiency represents a considerable weight reduction when compared to

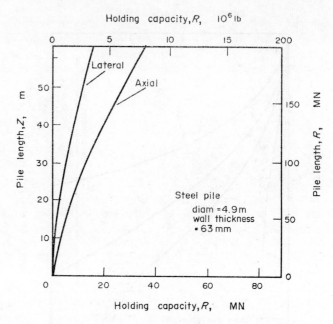

FIG. 12.    Comparison of axial (pullout) and lateral capacity of an anchor pile in a clay seafloor.

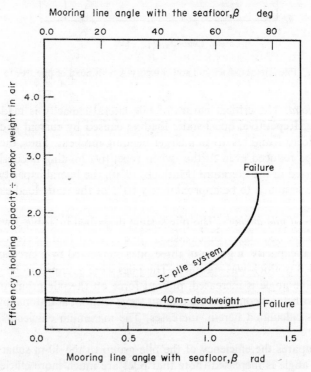

FIG. 13.    Effect of mooring line angle on efficiency of deadweight and pile anchors in clay seafloor.

the weight of the deadweight. However, for low line angles the efficiencies of piles and deadweight are about the same.

The pile system is adversely affected by, and highly susceptible to, failure under repetitive loading. At low line angles, pile system reliability is probably as good as deadweight system reliability since the weight of both systems is almost the same. At high line angles, however, the more efficient (lighter) pile system is probably less reliable. The pile group described in Table 3 was designed with a 40% reduction in static capacity to reduce the chance of failure due to repetitive loading.

Capacities plotted in Figs 10 and 11 are for a single pile with a free head. The effect of joining a number of these piles under a rigid pile cap was not considered. Pile groups for OTEC will probably have less axial capacity than the sum of individual pile capacities due to overlapping shear zones in the soil (Meyerhoff, 1976). Conversely, lateral capacity may actually be increased in a pile group (Kim and Brungraber, 1976; Bykov, 1975). Actual load tests at the proposed site or simulation of soil-pile interaction using finite element techniques will be required to confidently predict pile group behavior. At this stage of concept development the simple example given provides a conservative, yet representative, indication of pile anchor performance.

The primary difficulty with a pile anchor system is installation. First the ability to drill a 4.9-m-dia hole in 6,000 m of water has not been demonstrated. Second, a heavy lift system would be required to place the pile cap on the bottom. Third, the critical grouting operation would be especially tedious at these depths. Mooring line attachment and pile-to-cap bonding are two other complexities which must be considered.

The basic advantage of piles is their ability to efficiently resist pullout loads; their principal disadvantage is complex installation.

*Plate anchors*

*Summary of findings.* Figure 14 shows a typical plate anchor and its method of operation. The rectangular or square flat plate is driven deep into the soil by firing from a large-diameter gun. The plate is then rotated to the horizontal position by pulling on the eccentric mooring line attachment point.

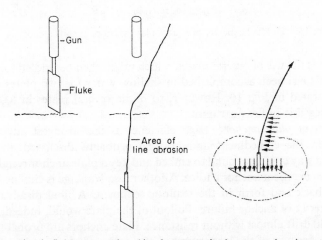

Fig. 14.   Simple fluke-type anchor (develops capacity by reverse bearing capacity).

Figure 15 illustrates the holding capacity of various square-plan plate anchors. The short-term holding capacity, indicated by the solid curves, governs realizable holding capacity. The probable maximum size of a plate anchor is limited by the ability to embed the plate. This ability appears limited in the near future to plate sizes about equal to the smallest one indicated in Fig. 15. This "small" plate would probably require a gun almost 490-mm (20 in.) inside diameter to drive it to the required embedment depth. Embedding the larger plates is probably not feasible using this method. Although other embedment methods are possible, none appears economically and technically justifiable in the near future.

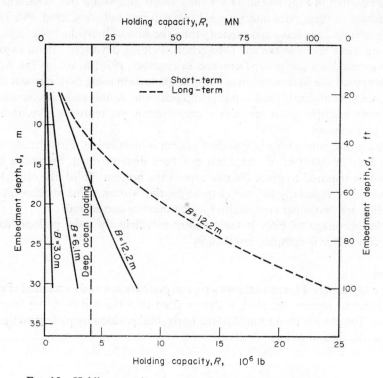

FIG. 15.   Holding capacity of square-plate anchors in soil category A.

An alternative is the use of several plates, which might then be bridled together to distribute the load. This has been accomplished in shallow water for two plates using a system like the one illustrated in Fig. 16. However, to bridle several plates in 6,000 m of water is probably not possible in the near term.

*Relative merits of plate anchors.* High efficiency is the strongest advantage of plate anchors. However, the embedment and bridling problems discussed above are major disadvantages. If a way could be found to embed and key a plate anchor roughly 12 × 12 m, it might become an excellent anchor choice. Another disadvantage is chafing of the mooring line as it works back and forth in the seafloor sediment. A final disadvantage of plate anchors is the severity of anchor failure. Pullout of the plate would, indeed, mean that the OTEC plant would drift almost without resistance. Plate anchors are probably not desirable as OTEC anchors because of the large number of significant negative factors.

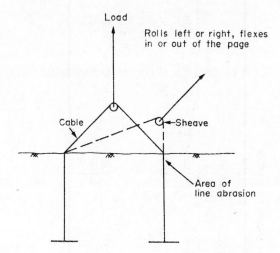

FIG. 16. Bridling technique for two plate anchors.

## Drag-embedment anchors

*Summary of findings.* Drag-embedment anchors are the anchors used routinely by ships and some large semisubmersibles. As shown in Fig. 17, the anchor is dropped to the seafloor, then forced to bury itself in the seafloor by pulling almost horizontally on the mooring line. The mooring line angle at the seafloor must be near zero in order for the anchor to embed. After the anchor is embedded then the mooring line angle with the seafloor can be increased to about 0.1 rad (6°) without substantially decreasing the ultimate holding capacity of the anchor. However, should the ultimate holding capacity be exceeded and the anchor be pulled up from its initial buried position, it will not re-embed itself unless the line angle is reduced to near zero. Because this form of anchor failure can be so drastic, the use of nonzero mooring line angles with drag-embedment anchors is not recommended. A typical drag-embedment anchor in widespread use today is shown in Fig. 18 and described in Town (1953).

Extrapolation of performance of existing drag-embedment anchors was done to determine whether or not they could be effectively used to moor OTEC. Table 4 presents the

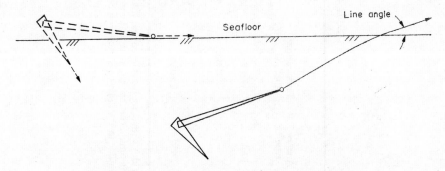

FIG. 17. Operation of the drag-embedment anchor.

Table 4. Summary of characteristics of possible drag-embedment anchors for OTEC

| Anchor type | Fabrication method | Maximum weight suggested possible, Mg (kips) | Holding capacity, MN ($10^6$ lb) | | Holding capacity cost, c/N (c/lb) | | Size (m) | |
|---|---|---|---|---|---|---|---|---|
| | | | Category A | Category D | Category A | Category D | Length | Width |
| Fluke | Welding | 68 (150) | 16.90 (3.8) | 22.24 (5) | 2.70 (12) | 2.02 (9) | 10.67 | 3.96*  11.58 |
| Fluke | Casting | 45.4 (100) | 11.12 (2.5) | 15.57 (3.5) | 1.80 (8) | 1.35 (6) | 9.14 | 3.35*  9.75 |
| Pick‡ | Casting | 63.5 (140)* | 15.12 (3.4)† | 21.35 (4.8)† | 1.80 (8) | 1.35 (6) | 10.36 | 3.96*  11.28 |
| | | 45.4 (100)† | 8.90 (2)† | 12.01 (2.7)† | 2.25 (10) | 1.69 (7.5) | 8.53 | 5.49 |
| | | 63.5 (140)† | 13.34 (3)† | 15.57 (3.5)† | 2.25 (10) | 1.80 (8) | 9.75 | 6.40 |

*Width with stabilizer hinged and folded.
†Maximum casting size, potentially involves quality control problems.
‡Similar to the fluke-type anchor in operation but cast as a single unit in a plow-like shape.

FIG. 18.   Danforth 2,510-lb (1,138 kg) anchor with fixed fluke.

results of that extrapolation. The cast fluke-type anchor was fairly representative of the extrapolated anchors and was chosen for comparison with other OTEC anchor candidates. Table 4 indicates that the probable maximum size for such an anchor is about 64 Mg $(140 \times 10^3 \text{ lb})$. It can be expected to produce a holding capacity of about 15 MN $(3.4 \times 10^6 \text{ lb})$.

Drag-embedment anchors have one other restriction which limits their usefulness in an omnidirectional loading environment. They resist load only in one general direction; load direction changes of over 1.57 rad (90°) would almost certainly cause the anchor to fail. Thus, even though a single anchor might provide the necessary holding capacity, multiple anchors would be required to resist omnidirectional loading.

*Relative merits of drag-embedment anchors.* Drag-embedment anchors, like plate anchors, exhibit a high holding-capacity-to-weight ratio. However, in order to realize this high efficiency, the use of drag-embedment anchors is limited to situations in which both the load direction and the mooring line angles are strictly limited. In the assumed deep ocean environment, these limitations mean that the drag-embedment anchor should be employed only in multipoint moors where a near-zero mooring line angle is maintained. Installation of a drag-embedment anchor mooring in deep water would be a complex task. A typical mooring might look similar to the eight-point version shown in Fig. 2. Each anchor would have to be embedded and set by pulling horizontally with a force on the order of 10 MN $(2.2 \times 10^6 \text{ lb})$. This operation would require a tensioning/mooring line stronger than any available today. Similarly, a powerful winch and a 10-MN $(2.2 \times 10^6 \text{ lb})$ reaction force would be required. The complex task of bridling anchors to equalize load would also be necessary. In view of the other options available, drag-embedment anchors are not attractive OTEC anchors.

### Summary of the comparison

Low line angles appear to be the most suitable for OTEC applications. Low line angles will minimize the already enormous mooring loads and will provide a more relaxed mooring, which will limit shock loads on the anchor and on the mooring line.

Deadweight anchors with skirts are by far superior for the deep ocean environment. They are reliable, relatively simple to install, and at low angles demonstrate an efficiency comparable with a pile anchor system.

Should line angles in excess of 0.79 rad (45°) be specified, the more efficient (less material weight) pile anchor system begins to appear more attractive. However, the savings in

TABLE 5.  RELATIVE MERITS OF SEVERAL OTEC GULF STREAM ANCHORS
(Horizontal load, 180 MN; water depth, 600 m.)

| Anchor Type | No. | Material | $\beta$* rad (deg) | Size, m | Weight in air, MN($10^6$ lb) | Reliability | Load direction capability | Repetitive load performance | Efficiency† | Hardware | Procedure | Comments |
|---|---|---|---|---|---|---|---|---|---|---|---|---|
| Deadweight | 1 | Unspecified | 0 (0)<br>0.79 (45) | 58 × 58 × 5.8<br>69 × 69 × 6.9 | 458 (102)<br>761 (171) | Good | Omni | Good | 0.39 | Heavy lift or buoyancy | Simple | Ability to place concrete under water needed; assumes bearing capacity controls. |
| Deadweight | 1 | Concrete | 0 (0)<br>0.79 (45) | 57 × 57 × 5.7<br>66 × 66 × 6.6 | 429 (96)<br>678 (152) | Good | Omni | Good | 0.43<br>0.37 | Heavy lift or buoyancy | Simple | Assumes bearing failure does not control; anchor design based on lateral load capacity. |
| Deadweight | 1 | Concrete | 0 (0)<br>0.79 (45) | 76 × 76 × 7.6<br>83 × 83 × 8.3 | 1,032 (232)<br>1,343 (302) | Good | Omni | Good | 0.17 | Heavy lift or buoyancy | Simple | Rock seafloor; ability to pour or place concrete under water needed. |
| Piles | 4<br>1<br>4<br>Total | Steel<br>Concrete cap<br>Grout | 0 (0)<br>0.79 (45) | 4.9 × 49 × 0.06<br>19.5 × 19.5 × 0.6<br>0.3 m (annulus) | 14.2 (3.2)<br>5.3 (1.2)<br>23 (5.1)<br>42.7 (9.6) | Fair | Omni | Poor | 4.2 | 4.9-m drill, heavy lift, grout | Intricate | Ability to drill, grout and emplace such large piles not demonstrated. |
| Piles | 1<br>144<br>Total | Steel<br>Concrete cap<br>Grout | 0 (0)<br>0.79 (45) | 0.25 × 20 (solid)<br>20 × 20 × 1<br>0.025 (annulus) | 6.6 (1.5)<br>9.4 (2.1)<br>1.5 (0.34)<br>17.6 (3.9) | Fair | Omni | Poor | 14.5 | Rock drill, heavy lift 1 × $10^6$ lb winch | Intricate | Rock seafloor; ability to drill and grout not demonstrated. |
| Plate | 12 | Steel | 0.79 (45) | 3.05 × 3.05 × 0.23 | 2.0 (0.44) | Fair | Omni | Fair | 128 | 4 × $10^6$ lb winch | Intricate | Bridling, embedding line chafing, keying. |
| Drag embedment | 9 | Steel (cast) | 0 (0) | 10 × 11 (fluke) | 5.6 (1.3) | Fair | Uni | Fair | 31 | | Intricate | Bridling, setting line chafing. |

*Mooring line angle with the seafloor.
†Efficiency = holding capacity ÷ weight in air.

materials may not be sufficient to offset the high installation costs and reduced reliability of a pile group.

Drag-embedment or plate-type anchors do not appear to be reasonable choices for OTEC anchors. Both suffer from line abrasion, bridling problems, and installation intricacies necessitated by the required multianchor moorings. In addition, reliability is low for both types. Drag anchors are even more severely limited than plate anchors because of the requirements for unidirectional loading and low line angles.

## COMPARISON OF ANCHOR TYPES FOR THE GULF STREAM ENVIRONMENT

The relative merits of the several anchor concepts considered are summarized in Table 5. The relative merits of piles vs deadweights in Gulf Stream areas are harder to distinguish than in the deep ocean environment. Much depends on the actual bottom composition at the OTEC site. However, the following information should at least indicate the probable weaknesses, problem areas, and limitations of each type of anchor.

### Assumed environmental characteristics

The Gulf Stream environment is a very high energy area when compared to the deep ocean. This environmental disadvantage is balanced somewhat by the advantages of being near land and in relatively shallow water. Its proximity to a high electricity-demand area facilitates use of a direct electric power cable to land. Also, repair and fabrication yards are close.

Table 6 summarizes the assumed maximum mooring forces and bottom characteristics in the Gulf Stream. While soil strength is generally greater (sand and an occasional rock bottom) and depths are decreased to about 600 m, mooring loads have increased by a factor of 10. At an assumed maximum line angle of 0.79 rad (45°), the maximum vertical force equals the maximum horizontal force (180 MN) 40 × 10⁶ lb); loading will probably be unidirectional.

TABLE 6.   LOADING AND CHARACTERISTICS OF THE GULF STREAM ENVIRONMENT

| Characteristics | Description |
|---|---|
| Sea condition | High energy |
| Depth | 600 m |
| Bottom soil | Categories D and E: sand and rock |
| Assumed upper limit* | 180 MN (40 × 10⁶ lb) |
| Range of mooring line angles considered | 0 to 0.79 rad (0 — 45°) |
| Vertical force | 0.0 to 180 MN (0.0 to 40 × 10⁶ lb) |
| Local direction | Unidirectional |

*Based on estimates of horizontal loading: 66.7 MN (15 × 10⁶ lb) and 70.7 MN (15.9 × 10⁶ lb).

### Comparison criteria

The criteria for comparing anchors in the Gulf Stream remain the same as those for comparing deep ocean anchors except for the following.

*Reliability.* Reliability is even more critical than with deep ocean anchors since shallow water and shipping traffic lanes are near the possible OTEC sites. The increased current velocities compound the problem by increasing the drift rate of the OTEC plant.

*Load direction.* The ability of an anchor to resist multidirectional loading will not be critical in this unidirectional load environment.

*Line angles.* The tremendous forces acting here again necessitate use of low mooring line angles, if at all possible. High line angles are more likely to cause liquefaction of the sediment types found in the Gulf Stream. Only line angles less than 0.79 rad (45°) are addressed here since higher line angles make the required line strengths untenable.

*Installation.* As mentioned previously, the advantages which accrue due to shallower operating depth are probably offset because of the increased magnitude of forces to be countered. Even without the large current forces, the assumed 600-m depth is somewhat beyond current offshore experience.

### Deadweight anchors

*Summary of findings.* The performance of deadweight anchors on a sand seafloor may be governed by either bearing capacity or lateral-load capacity.

That bearing capacity should be a problem is contrary to experience with building foundations on sand; normally, bearing capacity is not critical. The difference in predicted failure modes is due to the very high lateral-to-vertical-load-component ratios to which the soil under the deadweight anchor is subjected. A large lateral-load component increases the eccentricity and the inclination (to the vertical) of the resultant force on the supporting soil. Both factors, eccentricity and inclination, act to reduce the bearing capacity of a deadweight. The practical effect of eccentricity is to concentrate the load over a smaller area of the supporting soil; the practical effect of inclination is to shift the potential failure surface beneath the deadweight. This tends to decrease the normal load acting on that surface and thus to reduce the resistance to load. Increased eccentricity and increased inclination both act to reduce the load that can be supported by the soil.

In sand, the bearing capacity factors $i_q$ and $i_r$ (Fig. 19) reflect this apparent decrease in bearing capacity due to inclination alone. Note that as the ratio of horizontal-to-vertical load (effective weight) approaches unity, these factors (and therefore bearing capacity)

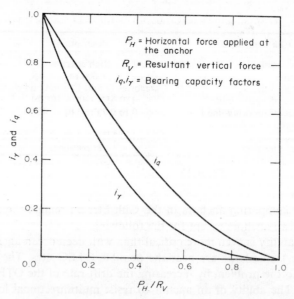

FIG. 19.   Bearing-capacity coefficients for inclined loads in soil category D.

approach zero. To increase these factors the ratio of horizontal-to-vertical load must be decreased, but horizontal load is set by environmental conditions. The only alternative is to increase the downward vertical-load component by increasing the mass of the anchor block. This decreases the inclination of the resultant. The ratio of horizontal-to-vertical load required to prevent a bearing failure is 0.65. Simply stated, the effective submerged weight on the seafloor must be at least 1.5 times as great as the horizontal load to prevent the bearing-capacity-type failure of the soil. Figure 20 illustrates the weight required to prevent bearing failure ($R_R = Q$-curve) for a line angle of zero. The equation provided on the figure specifies required weights for line angles other than zero.

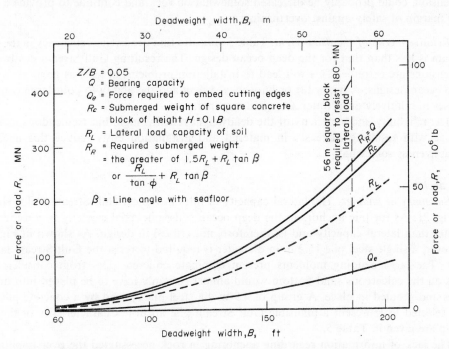

FIG. 20.   Relation of deadweight bearing capacity to various other loadings in soil category D.

The weight required in a deadweight anchor may be governed instead by the lateral-load capacity required. The lateral-load capacity is directly proportional to the combined weight of the deadweight and the soil included within the skirts or cutting edges. Table 5 includes the description of a deadweight designed specifically to resist sliding under the 180-MN ($40 \times 10^6$ lb) lateral load. For the configuration chosen and for the site conditions specified, the deadweight submerged weight required to resist sliding is 246 MN ($55 \times 10^6$ lb) (429 MN air weight, if concrete ($96 \times 10^6$ lb)), while the submerged weight required to prevent local bearing failure is 274 MN ($60 \times 10^5$ lb) (458 MN air weight ($103 \times 10^6$ lb))—only slightly higher. The deadweight would presumably fail in a local bearing mode, tilting toward and sinking on its leading edge.

The practical significance of such a local-bearing failure has not been fully investigated. The hazard associated with a bearing failure of this type is a tilting of the anchor block. When the load direction changes, this tilting could lead to line chafing on the high side of the anchor. In a unidirectional loading environment, this type of failure could be inconsequential. In any event, the difference in calculated weights required for bearing capacity, compared to the weight required for lateral-load capacity (274 MN vs 246 MN), is small for the boundary conditions given.

The design of a deadweight for a rock seafloor was straightforward. Strict frictional behaviour of rock against concrete was assumed to provide lateral resistance; a friction coefficient of 0.3 was used. If loose rock fragments or unconsolidated sediments are present, the friction coefficient could be as low as 0.1. The deadweight was sized to provide the required normal force. Deadweight height, assumed to be equal to 0.1 times the base dimension, could probably be increased somewhat on rock and continue to provide a suitable margin of safety against overturning.

*Relative merits of the deadweight.* Table 5 indicates that deadweight efficiency in the Gulf Stream is less than that for the deep ocean design. The resulting Gulf Stream deadweight dimensions are extreme. This will lead to installation problems at least as great as for the deep ocean designs. A heavy lift or buoyancy system and the ability to pour large concrete masses in relatively deep water are again required.

The reliability and simplicity of the deadweight remains high, but tremendous increase in size with resulting increases in material and installation costs reduce this anchor's attractiveness somewhat.

### Pile anchors

*Summary of findings.* The lateral capacity of piles in typical Gulf Stream soil is shown in Fig. 21. As for pile anchors in the deep ocean soils, pile axial capacity is considerably greater than lateral capacity and is, therefore, not critical in design. As shown in Fig. 21. however, a single steel pile 15.2 m in diameter is required to resist the Gulf Stream lateral load. The large bending moments present eliminate concrete piles from consideration, Piles on the calcareous sand bottom would almost certainly have to be placed into drilled holes and grouted in place. A group of 4.9-m-dia steel piles joined by a concrete pile cap was selected to represent a plausible Gulf Stream pile anchor. Characteristics of the pile group are given in Table 5.

The lack of information regarding anchoring in rock necessitated the oversimplified— yet conservative—design specified in Table 5. Assumptions and method were outlined in the second progress report (Valent *et al.*, 1976b).

*Relative merits of pile anchors.* Table 5 clearly shows the advantage offered by pile anchors. The mass of materials required is only 6% as much as for a deadweight on sand (line angle = 0.79 rad) and only 1.3% of that required for a deadweight on rock. Pile anchor reliability and reaction to repetitive loading remain about the same in the Gulf Stream environment as in the deep ocean environment.

The complexity of installing a pile anchor group remains a major disadvantage in using this anchor system. Note that a lift system capable of lowering 5 MN (1.2 × 10⁶ lb) is needed to place the pile cap.

The ability to drill large-diameter holes (4.9 m) at a water depth of 600 m, to install the large-diameter steel piles, to grout them in place, and to connect the piles using a rigid pile cap represents a complex sequence of tasks which must be mastered before a pile anchor can be installed on a sand seafloor. For a rock seafloor, the ability to drill in rock in great water depths must be developed. However, if offshore technology can be quickly

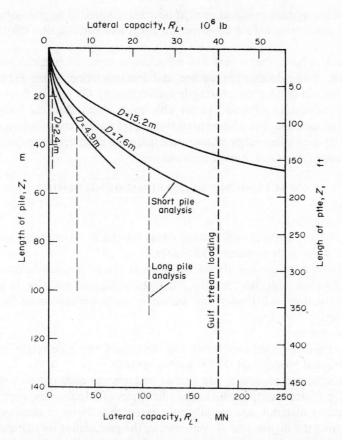

FIG. 21.   Pile lateral capacity in soil category D with head of pile free to rotate.

extended to accomplish these tasks, piles may surpass deadweights as the optimum OTEC anchor in the Gulf Stream.

### Plate anchors

Plate anchors of the size and number required for the Gulf Stream environment are considered an unreasonable concept. As shown on Table 5, twelve 3-m, square-plate anchors, driven to a depth of 30 m would be required. Embedding, setting, bridling, and preventing line chafe are all serious problems.

### Drag-embedment anchors

As with plate anchors, the size and number of standard burial anchors required for anchoring OTEC in the Gulf Stream environment is considered unrealistic. Fabrication, installation, and line chafing would remain significant problems.

### Summary of the comparison

Both deadweight and pile anchor systems offer a viable solution to the problem of anchoring the OTEC platform in the Gulf Stream environment. Deadweight anchors are the favoured anchor system choice at present because (compared to pile systems) the deadweight requires much less seafloor construction effort and because that effort is not nearly as complex.

However, deadweights offer a very low efficiency in terms of material weight and cost. Also, fabrication of the deadweight anchor and transportation to the OTEC site would be more expensive and difficult than for a pile anchor system. On rock seafloors, pile anchors offer the added advantage of being better able to accommodate the irregular seafloor surface. Thus, the development of improved techniques for the construction of pile groups in deep water may shift pile anchor systems, compared to deadweight anchors, into a more favourable position for a Gulf Stream environment.

## CONCLUSIONS AND RECOMMENDATIONS

*Anchor selection*

*Deep ocean environment.*

1. The deadweight anchor with cutting edges (skirts) is an effective and economical anchor system for use at deep ocean OTEC sites.

2. Pile anchor systems are less attractive for use in the deep ocean because a significant amount of seafloor construction time for pile installation and grouting is required. This at-sea construction time will render pile anchor systems uneconomical for use at deep ocean OTEC sites.

*Gulf Stream environment.*

1. On unconsolidated seafloors (sand and stiff clay) the deadweight anchor system offers an effective and economical OTEC anchor system.

2. On unconsolidated seafloors, pile anchor systems are competitive with deadweight systems in the Gulf Stream environment. The efficiencies of pile anchors, expressed in terms of weight of anchor material, are considerably better than those of deadweight anchors, somewhat offsetting the higher cost of constructing the pile anchor on the seafloor.

3. On rock seafloors, the pile anchor system offers considerable savings in material over the deadweight anchor (provided that a suitable installation system can be developed).

*Technological refinements required*

*Deep ocean environment.*

1. Utilization of the deadweight anchor concept will require:

a. Either a heavy lift or controlled buoyancy system capable of handling and lowering to 6000 m weights approaching 18 MN ($4 \times 10^6$ lb) submerged;

b. Or a lift or controlled buoyancy system capable of handling and lowering to 6,000 m a frame (or box) weighing 1.8 MN ($0.4 \times 10^6$ lb) submerged, coupled with a weighting technique or system capable of adding the remaining 16 MN ($3.6 \times 10^6$ lb) submerged, to the frame (the weighting could be any suitable material such as large concrete blocks, stone, iron ore, poured-in-place concrete, or even seafloor sediment);

2. Utilization of the pile anchor concept will require capabilities similar to those required for deadweight installation. In addition, the pile anchor concept will require development of a grouting technique or system for 6,000 m to cement the piles into calcareous-ooze seafloors and to cement the pile heads into the pile cap.

*Gulf Stream environment.*

1. Utilization of the deadweight anchor concept will require:

a. Either a controlled buoyancy system capable of handling and lowering to 600 m on unconsolidated seafloors, weights approaching 260 MN ($60 \times 10^6$ lb) submerged, or on rock seafloor weights to 600 MN ($133 \times 10^6$ lb) submerged;

b. Or a heavy lift or a controlled buoyancy system capable of handling and lowering to 600 m a frame (or box) approaching 60 MN ($13 \times 10^6$ lb) submerged, coupled with a technique or system for adding 540 MN ($120 \times 10^6$ lb) of weighting to that frame.

2. Utilization of the pile anchor concept will require:

a. A heavy lift or a controlled buoyancy system capable of handling and lowering the pile cap to the seafloor to a water depth of 600 m. That pile cap may weigh 9 MN ($2 \times 10^6$ lb) submerged;

b. A pile anchor installation system capable of drilling large diameter holes (to 4.9 m) in calcareous sands and rock, of lowering steel pipe piles into these holes, and of cementing these steel piles into sediments and rock.

*Acknowledgement*—We gratefully acknowledge the assistance and constructive criticism of many colleagues at CEL and others associated with the OTEC project. Direct CEL contributors to the effort included R. M. Beard on plate anchor holding capacities, H. J. Lee on the soil categories, J. F. Wadsworth on anchoring on rock, and R. D. Rail on the anchor pile structural designs. Credit for much of the calculation, tabulation and drawing preparation belongs to L. J. Wolosynski, F. O. Lenhardt and P. H. Babineau, all of CEL. Last, but not least, the authors thank their report reviewers for helping in producing a more useful work: R. A. Breckenridge, CEL; CDR Larry Donovan and Gene Silva, NAVFAC, Alexandria, Virginia; and Hal Skowbo, ERDA, Washington, D.C.

## REFERENCES

BYKOV, V. I. 1975. Experimental performance of horizontally loaded pile foundations. *Soil Mech. Foundation Engng* **12**, 103.

DOUGLASS, R. H. 1975. Ocean thermal energy conversion: An engineering evaluation, in *Proc. Third Workshop Ocean Therm. Energy Conversion (OTEC), Houston*, p. 28, Laurel, MD, Johns Hopkins University, Applied Physics Laboratory.

KIM, J. B. and BRUNGRABER, R. J. 1976. Full-scale lateral load tests of pile groups, *J. Geotec. Engng Div. ASCE* **102**, 87–105.

LITTLE, T. E., MARKS, J. D. and WELLMAN, K. H. 1976 Deep water pipe, pump, and mooring study, Ocean Thermal Energy Conversion Program, Final Report, U.S. Energy Research and Development Administration. Annapolis, MD, by Westinghouse Electric Corporation, Oceanic Division, June 1976, pp. 5–101. (Contract No. (11-1)-2642).

MATLOCK, H. 1970. Correlations for design of laterally loaded piles in soft clay, in *Proc. 2nd Ann. Offshore Technol. Conf.* Houston, OTC Paper 1204.

MEYERHOF, G. G. 1976. Bearing capacity and settlement of pile foundations, *J. Geotec. Engng Div. ASCE*, vol. **102**, no. GT3, Mar 1976, pp. 197–225.

TOWNE, R. C. 1953. Test of anchors for moorings and ground tackle design, Naval Civil Engineering Laboratory, Technical Memorandum 066. Port Hueneme, CA, 10 Jun 1953.

VALENT, P. J. *et al.* 1976a. State-of-the-art in high-capacity, deep-water anchor systems, Civil Engineering Laboratory, Technical Memorandum 42-76-1. Port Hueneme, CA, Jan 1976.

VALENT, P. J. *et al.* 1976b. Holding capacities of single anchors in typical deep-sea sediments, Civil Laboratory, Technical Note (in publication). Port Hueneme, CA. 1976.

*Ocean Engng.* Vol. 6, pp. 169-245. Pergamon Press 1979. Printed in Great Britain

# SINGLE ANCHOR HOLDING CAPACITIES FOR OCEAN THERMAL ENERGY CONVERSION (OTEC) IN TYPICAL DEEP SEA SEDIMENTS

P. J. Valent, R. J. Taylor, J. M. Atturio and R. M. Beard

Civil Engineering Laboratory, Naval Construction Battalion Center, Port Hueneme, California 93043, U.S.A.

## NOMENCLATURE

| | |
|---|---|
| $A$ | Bearing area of deadweight anchor (m²) or cutting edge tips |
| $A_s$ | Side area (m²) |
| $B$ | Width of deadweight anchor block (m) |
| $b$ | Spacing between cutting edges beneath deadweight, or anchor spacing (m) |
| $D$ | Diameter of pile or circular anchor block (m) |
| $d$ | Depth below seafloor surface (m) |
| $d_c, d_\gamma, d_q$ | Factors to account for the shearing resistance of soil above the bearing plane of the anchor |
| $E$ | Young's modulus (Pa) |
| $e$ | Distance of resultant force from centroid of base (m) |
| $H$ | Height of deadweight anchor excluding cutting edges (m) |
| $I$ | Polar moment of inertia of the pile (m⁴) |
| $i_c, i_\gamma, i_q$ | Factors to account for inclination of the resultant load |
| $K$ | Modulus of horizontal subgrade reaction (Pa) |
| $K_p$ | Coefficient of passive lateral earth pressure |
| $K_s$ | Coefficient of lateral pressure on pile wall |
| $k_c, k_\gamma, k_q$ | Factors to account for compressibility of the support soil |
| $L$ | Horizontal length, as the long dimension of a rectangular anchor block (m) |
| $M$ | Moment (Nm) |
| $M_{max}$ | Maximum moment (Nm) |
| $M_o$ | Moment at surface (Nm) |
| $M_p^1$ | Maximum bending moment in a cutting edge per unit width of deadweight anchor (Nm/m) |
| $M_z$ | Moment at a point (Nm) |
| $N_c, N_\gamma, N_q$ | Bearing capacity factors |
| $n_h$ | Coefficient of horizontal subgrade modulus (Pa/m) |
| $P$ | Force applied to an anchor by the mooring line (N) |
| $P_H$ | Horizontal component of mooring line force at the anchor (N) |
| $P_V$ | Vertical component of mooring line force at the anchor (N) |
| $p$ | Soil pressure (Pa) |
| $Q$ | Vertical load reaction supplied by soil (N) |
| $Q_e$ | Force required to embed cutting edges (N) |
| $Q_s$ | Total side friction on pile (N) |
| $R$ | Anchor capacity (N) |
| $R_A$ | Axial force capacity of pile (N) |
| $R_b$ | Lateral load resistance developed below the cutting edges of the deadweight, base shear resistance (N) |
| $R_G$ | Submerged weight of concrete deadweight anchor (N) |
| $R_L$ | Total lateral resistance or capacity (N) |
| $R_p$ | Lateral load resistance of passive wedge against leading cutting edge of deadweight (N) |
| $R_p^1$ | Lateral load resistance per unit width of passive wedge (N/m) |
| $R_R$ | Required submerged weight (N) |
| $R_V$ | Net vertical force (N) |
| $S_p$ | Bearing capacity shape factor |
| $S_t$ | Sensitivity = (undisturbed undrained strength) / (remolded undrained strength) |
| $s$ | Undrained shear strength (Pa) |

| $s_c, s_\gamma, s_q$ | Factors to account for base shape |
|---|---|
| $s_f$ | Friction against cutting edge lateral area (Pa) |
| $C_m$ | Average undrained shear strength (Pa) |
| $T$ | Relative stiffness (m) |
| $t$ | Thickness of cutting edge (M) |
| $t_w$ | Thickness of pile wall (m) |
| $V$ | Shear force (N) |
| $W$ | Weight (N) |
| $W_b$ | Submerged weight (N) |
| $W_{eff}$ | The effective weight or force on the base shear plane (N) |
| $W_s$ | Submerged weight of soil entrained within the cutting edges and above the base shear plane (N) |
| $x$ | Distance along pile below point of lateral load application (m) |
| $y$ | Lateral deflection (m) |
| $Z$ | Length of structural element (m) |
| $\beta$ | Angle of mooring line with horizontal (rad) |
| $\gamma$ | Buoyant unit weight (N/m³) |
| $\delta$ | Effective friction angle of soil on pile wall or cutting edge wall (rad) |
| $\lambda$ | Friction capacity coefficient |
| $\bar{\sigma}$ | Vertical effective stress (Pa) |
| $\bar{\sigma}_m$ | Average vertical effective stress over pile length (Pa) |
| $\bar{\varphi}$ | Effective angle of internal friction (rad) |

## INTRODUCTION

*Purpose*

This article presents performance characteristics of possible anchors for the Ocean Thermal Energy Conversion power plant (OTEC). The first report (Valent *et al.*, 1976) gave a general description of possible OTEC anchor concepts.

*Procedure*

Each anchor type presented in the first OTEC anchor report was evaluated using available data and standard analysis techniques. All calculations were made assuming that load was statically applied. The advantages and limitations of combining several anchors of the same type to achieve desired capacity were discussed.

Anchor holding capacities for OTEC were calculated for the soil categories described in the first report. A summary of these soil categories and assumed anchor loading is given in the next section.

## ENVIRONMENT

*Site categories*

The sediment strength profiles described in the first report were grouped into three site categories. One shear strength profile was developed to represent each category (Fig. 1). These three site categories, plus two additional categories for sands and for outcropping and near-surface seafloor rock respectively, were used as representative OTEC seafloor environments. The selection, sizing, and comparing of anchor types and combinations refers back to these baseline categories.

Table 1 is a summary listing of site category characteristics. Note that each category has been assigned a descriptive word title in Table 1; for example, the sediments of site category A can be descriptively referred to as clays and clayey silts.

For site category D, sands, a predominantly quartz grain material is assumed having a friction angle of 0.52 rad (30°) and a submerged unit weight of 6.8 kN/m³ (43 pcf). Such material under repeated loading may be susceptible to liquefaction. If the median grain

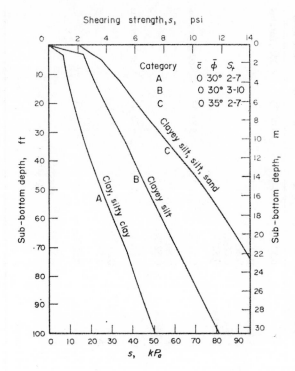

FIG. 1.   Properties assumed for 3 of 5 assumed seafloor categories.

size ($D_{50}$) is found to lie between 0.02 and 0.2 mm, then the mooring system should be designed to reduce the effects of repeated loading or high factors of safety (greater than 10) should be used.

The material characteristics of site category E, that for exposed or shallowly buried rock, are not resolved at this point. The quantified rock properties are not nearly as significant in comparing anchor types as are the mechanisms for engaging the rock surface and the techniques for installing the anchor system. Seafloor rock, category E of Table 1, will be filled in as the data need develops.

For each of the site categories, the performance of the various anchor types was determined. Thereby, those anchor types best suited for a given environment (and loading situation) were identified.

*Loading*

*Horizontal.* Given input information from some of the OTEC power plant proponents/ concepts, CEL established upper limits to the range of horizontal load magnitudes to be considered in its anchor study. Because the present designs are all preliminary and because this study is to encompass all possible candidate power plant systems, the upper bound to required horizontal holding capacity was set at about twice the capacities required for present plant concepts. For the deep ocean site, static drag forces of 6 MN (1.27 × 10⁶ lbs) and 9 MN (2 × 10⁶ lbs) were estimated for two concepts: for such a site, the upper bound of the horizontal load capacity range for this study was set at 18 MN (4 × 10⁶ lbs). For

Table 1. Measured/calculated profile groupings and assumed site category characteristics

| Category | Profile* | Figure* | Description | s (kPa/psi) | $\bar{\varphi}$(rad/deg.) | $S_t$ | $\gamma$ (kN/m³/pcf) | Comments |
|---|---|---|---|---|---|---|---|---|
| A | | | Clay; Silty clay† | 0 | 0.52/30 | 2–7 | 4.08/26 | <25% forams |
| | 1 | 2 | Pelagic clay, abyssal hill province | 3.45/0.5 | 0.61/35 | 5–7‡ | | |
| | 2 | 3 | Clay and nannofossil calcareous ooze | 0 | 0.66/37 | 5 | | |
| | 3 | 5 | Soft basin material | 1.38/0.2 | 0.66/37 | 2–4 | | |
| | 4 | 4 | Distal turbidite (weak) | 2.76/0.4 | 0.52/30 | 3–5 | | |
| B | | | Clayey silt† | 0 | 0.52/30 | 3–10 | 5.81/37 | >60% forams |
| | 1 | 3 | Loose foram ooze, calcareous | 0 | 0.58/33 | 10 | | |
| | 2 | 4 | Distal turbidite (strong) | 2.76/0.4 | 0.52/30 | 3–5 | | |
| C | | | Clayey silt, silt† | 0 | 0.61/35 | 2–7 | 6.75/43 | >60% forams |
| | 1 | 4 | Proximal turbidite (silt) | 1.72/0.25 | 0.68/39 | 2–3 | | |
| | 2 | 3 | Dense foram ooze, calcareous | 0 | 0.58/33 | 5–7 | | |
| D | | | Sand† | 0 | 0.52/30 | 10 | 6.75/43 | |
| | 1 | 4 | Proximal turbidite (sand) | 0 | 0.52/30 | potentially liquefiable | | |
| E | | | Seafloor rock, exposed or shallowly buried† | | | | | |

*For data source refer to Valent et al., 1976.
†Values are based on engineering judgments.
‡For upper 4.6 m; 3 to 5 for below 4.6 m.

the Gulf Stream environment, the horizontal loading at the anchor was estimated at 67 MN to 71 MN (15 × 10⁶ lbs) and (15.9 × 10⁶ lbs) for two concepts: for the Gulf Stream site, the upper bound of the horizontal load capacity range has been set at 180 MN (40 × 10⁶ lbs). Note that these load magnitudes are steady state due to wind and current drag forces. The load component at the anchor due to wave action on the floating structure is assumed to be completely damped out by the mooring line system.

*Vertical.* At the beginning of the CEL OTEC anchor effort a wide spectrum of mooring line angles and vertical mooring line force components was considered (Fig. 2). Two loading envelopes were assumed. The first, representing the deep ocean, "benign" environment, assumed a maximum mooring line angle with the horizontal, β, of 1.40 rad (80°) with a corresponding maximum vertical load component of 100 MN (22.7 × 10⁶ lbs). The high line angle of 1.40 rad was selected to include a proposed buoyant line design in which the deep section of the mooring line was to be highly positively buoyant rising at a steep angle from the anchor. The second loading envelope, representing the Gulf Stream environment, assumed a maximum mooring line angle of 0.79 rad (45°) with a corresponding maximum vertical load component of 180 MN (40 × 10⁶ lbs). Reference to these loading envelopes will be made throughout this report.

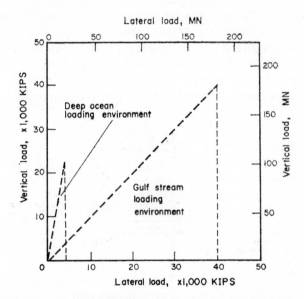

FIG. 2.   Assumed envelope of mooring line forces at the anchor (Jan. 1976).

The authors do make note here that, at the time of writing, the high mooring line angles and the corresponding high vertical mooring line force components at the anchor do not appear practical. Perhaps the most significant drawback to the high line angles in a conventional mooring would be the increased stiffness of the moor and the resulting high dynamic load carried in the mooring line and transmitted to the anchor and supporting soil. Such high dynamic load acting on the predominantly calcareous oozes of the potential OTEC siting zones will result in significant soil holding capacity reductions. This necessitates increased anchor dimensions to reduce the critical stress levels. A better solution to the dynamic loading problem would seem to be to reduce the dynamic stress level by making the mooring more compliant by lowering the line angle.

## STANDARD BURIAL ANCHORS

### Introduction

Standard burial anchors are a potential choice for anchoring structures of the sizes projected for OTEC. These burial anchors do not exist, however, in sizes large enough to yield holding capacities commensurate with simplified installation and efficient use of components. Use of the existing small anchors would result in very difficult load equalization problems. Overdesign would be required due to the high degree of uncertainty involved in achieving load equalization.

The largest anchor advertised available weighs 45 Mg (100,000 lbs); the largest anchor tested weighs 13.6 Mg (30,000 lbs); while the largest anchor tested sufficiently to yield data that could be used to extrapolate to sizes needed for OTEC was 6.8 Mg (15,000 lbs). This is a long way from the anchor weight required to yield holding capacities in the 13.3 MN to 44.5 MN ($3 \times 10^6$ to $10 \times 10^6$ lb) range, the range which might be practicable.

### Performance prediction for scaled-up anchors

Three basic types of burial anchors were chosen for extrapolation, the standard fluked type (e.g. STATO, DANFORTH, BOSS), the pick type (e.g. BRUCE, HOOK, ADMIRALTY MOORING), and the mud type (e.g. DORIS, PARAVANE) anchors. Using relationships between anchor weight and anchor holding capacity (Valent *et al.*, 1976) the required weights of these anchors could be determined for various soils and holding capacities. Extrapolation can be accomplished only for: (1) geometrically similar anchors,

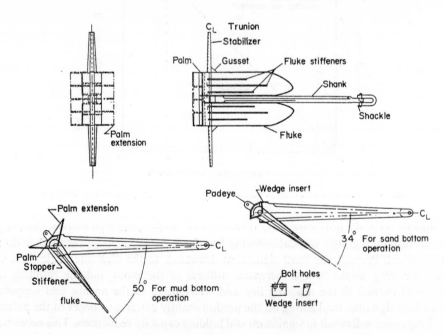

Fig. 3.   NAVFAC STATO mooring anchor (from Anon., 1968).

or (2) varying anchor geometries where the effect of variation on holding capacity is known. In order to provide some measure of confidence in the results of the extrapolation, geometric similarity will be maintained.

*Fluked type anchor.* The STATO configuration (similar to OFFDRILL, MOORFAST, STAYRITE) will be scaled-up because the majority of available data exists for this particular anchor. Figure 3 details the characteristics of the STATO anchor. If the existing STATO anchors were geometrically similar, the efficiency of the anchor type would be constant (Coombes, 1931). The data show, however, that efficiency decreases slightly with size. This occurs because the linear dimensions of each anchor do not vary exactly as weight to the one-third power, i.e. $W^{1/3}$. The requirement that mild steel be used for fabrication caused plate thickness to increase as about $W^{1/2}$ for the larger sizes. For purposes of this evaluation, the 5.5 Mg (12,000 lb) STATO, which exhibits efficiencies at maximum drags of 25/1 in soft clay and 35/1 in sand, will be used as the basis for this extrapolation. The latter value was controlled by steel stress rather than soil strength.

Geometric similarity is maintained by increasing steel strengths as well as the linear dimensions in proportion to $W^{1/3}$. Characteristics of the scaled-up anchors are given in Table 2. Anchor weights required to produce these capacities are shown in Fig. 4. The required weight in the lower capacities is dependent upon soil strength; whereas, anchor steel stress controls weight in the larger capacities. It is apparent from Fig. 4 that once the criterion of limiting steel stress is achieved, anchoring efficiency no longer remains constant. Data from tests on existing anchors indicate that holding capacity increases in proportion to $W^{0.69}$. This is extremely close to the theoretical limit of $W^{2/3}$. This slight difference is attributed to errors in transferring data from log-log plots of actual steel stress vs anchor

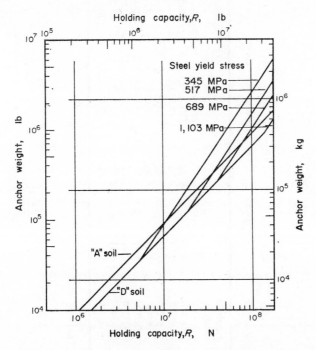

FIG. 4.  STATO anchor weights required to produce desired holding capacities.

TABLE 2. Characteristics of existing and extrapolated STATO anchors (refer to Fig. 3)

| Nominal weight | Anchor dimensions in meters | | | | | | | |
| | Fluke | | | Shank | | Palm extension | | Stabilizer |
| Mg/kips | Length | Width | Thickness | Length | Width | Length | Width | Length |
|---|---|---|---|---|---|---|---|---|
| 5/12* | 2.74 | 0.76 | 0.051 | 4.72 | 0.015 | 1.77 | 0.46 | 4.91 |
| 9/20 | 3.26 | 0.91 | 0.061 | 5.61 | 0.018 | 2.07 | 0.55 | 5.94 |
| 18/40 | 4.08 | 1.16 | 0.076 | 7.07 | 0.023 | 2.62 | 0.70 | 7.47 |
| 54/120 | 5.91 | 1.65 | 0.109 | 10.15 | 0.033 | 3.78 | 1.01 | 10.76 |
| 91/200 | 7.01 | 1.95 | 0.130 | 12.07 | 0.038 | 4.48 | 1.16 | 12.77 |
| 181/400 | 8.84 | 2.47 | 0.163 | 15.21 | 0.048 | 5.64 | 1.46 | 16.09 |
| 454/1000 | 11.98 | 3.32 | 0.221 | 20.63 | 0.066 | 7.65 | 1.98 | 21.85 |

*Existing anchor.

weight. Based upon this behavior, the most efficient anchor occurs at the intersection of the curves plotted for soil type and steel yield stress.

Another anchor, included in this category but somewhat different than the movable fluke type previously extrapolated, is the BOSS anchor, a fixed solid fluked anchor. Existing data show this to be a very efficient anchor in small sizes; however, there was insufficient data to confidently extrapolate performance. A conservative approximation of holding capacity was thus proposed.

$$R = 610 \ W^{0.75}$$

where 
$$R = \text{holding capacity (N)} \tag{1}$$
$$W = \text{weight (N)}.$$

The exponent "0.75" was recommended by Cole and Beck (1969). The extremely high efficiencies in the small sizes are attributed to the thin, high strength sections used in the anchor. Without modifying the design by adding stiffeners on the flukes, the practical limit of this design, maintaining geometric similarity, is probably much less than 45 Mg (100,000 lbs). By adding stiffeners to this anchor, the performance should more closely resemble STATO type designs since fluke/shank angles and effective fluke areas are approximately similar. A primary advantage of the BOSS anchor is its ability to embed in competent seafloors, a problem often encountered with other fluked anchors. The technique of fixing the fluke on one side of the shank may be appropriate for the movable fluke anchors, if they are used in competent seafloors.

A pertinent characteristic of the enlarged fluke anchors is their expected penetration depth. This information is needed to enable design of the anchor pendant in contact with the seafloor and provide an idea of how much lateral movement is necessary to fully set the anchor. A reasonable approximation in clay is that the setting distance to achieve peak capacity is six times penetration depth. For example, penetration depths in soil category A are found as follows:

(1) Computed required soil strength using anchor dimensions (Table 2) and peak design capacity.
(2) Obtain depth for required soil strength from shear strength vs depth curves (Valent et al., 1976).
(3) Plot depth vs anchor weight (see Fig. 5).

The required penetration of a 45 Mg (100,000 lb) anchor would be 33.5m (110 ft) and total setting distance would be about 201 m (660 ft). Also plotted in Fig. 5 are the actual penetration data for the STATO anchors in San Francisco Bay mud; the strength of this mud is very similar to category A soil. The remarkable agreement between actual and predicted penetration supports the extrapolated curve.

*Pick type anchor.* The BRUCE and HOOK anchors are pick type anchors. Relatively recent developments, they appear equally efficient (about 20/1 in category A soil and 28/1 in category D soil).

More data are available for the BRUCE anchor and it will, therefore, be used for extrapolation. The BRUCE anchor has one inherent advantage over the HOOK anchor in that it will always embed when pulled laterally, even if it lands on the shank side during deployment. This greatly simplifies deep water deployment.

Characteristics of the BRUCE anchor taken from company literature (Bruce Ltd., 1974) are presented in Table 3 along with those of several extrapolated anchors to

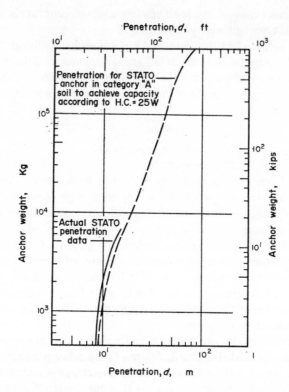

FIG. 5.   Predicted penetration of the STATO type anchor vs anchor weight.

455 Mg ($1 \times 10^6$ lb) size. The 455 Mg anchor, if it could be fabricated, would yield 89 MN ($20 \times 10^6$ lbs) holding capacity. The limiting characteristic, as with the STATO type, will be steel stress. Maximum holding capacity is plotted vs anchor weight for category A and D soils and for various limiting steel stresses in Fig. 6. The steel currently used for the BRUCE anchor has a yield stress of 500 MPa (72.5 ksi) which is greater than that required for the existing 6.4 Mg (14,000 lb) anchor. Maximum stress at maximum holding capacity for the 6.4 Mg anchor in the shank is 21 MPa (30 ksi) in soft clay (category A) and 29 MPa (42 ksi) in sand (category D). From Fig. 6, assuming a construction capability exists, it is apparent that approximately a 17.8 MN ($4 \times 10^6$ lb) capacity 91 Mg (200,000 lb) anchor could be fabricated using exactly the same steel that is used in the 6.4 Mg (14,000 lb) anchor by simply scaling-up the dimensions in proportion to weight to the one-third power, i.e. $W^{1/3}$.

Also included in Fig. 6 is a curve representing the theoretical relationship between holding capacity and anchor weight in that regime where steel stress control efficiency, i.e. holding capacity $\simeq$ (weight)$^{2/3}$. The initial point of the curve came from the company advertised limiting load of 2.98 MN (670,000 lbs) for the 6.4 Mg anchor fabricated from 500 MPa steel.

*Mud type anchor.* Only two of this type anchor are known to exist and are called the PARAVANE anchor and the DORIS Mud anchor. The existing sizes and company

TABLE 3.  CHARACTERISTICS OF EXTRAPOLATED BRUCE ANCHORS

| Nominal wt Mg/kips | A(m) | B(m) | C(m) | D(m) | E(m) | F(m) | G(m) | H(m) | J(m) | R(m) | RR(m) |
|---|---|---|---|---|---|---|---|---|---|---|---|
| 6.2/14 | 2.56 | 4.54 | 2.90 | 0.10 | 0.20 | 0.67 | 0.53 | 0.12 | 0.09 | 0.38 | 0.14 |
| 9.1/20 | 2.87 | 5.12 | 3.26 | 0.11 | 0.23 | 0.75 | 0.60 | 0.13 | 0.10 | 0.43 | 0.15 |
| 18.1/40 | 3.63 | 6.43 | 4.11 | 0.14 | 0.29 | 0.94 | 0.75 | 0.17 | 0.13 | 0.54 | 0.19 |
| 54.4/120 | 5.21 | 9.30 | 5.91 | 0.20 | 0.41 | 1.36 | 1.09 | 0.24 | 0.18 | 0.73 | 0.28 |
| 90.7/200 | 6.19 | 11.00 | 7.01 | 0.24 | 0.49 | 1.62 | 1.29 | 0.29 | 0.22 | 0.92 | 0.33 |
| 131.4/400 | 7.80 | 13.90 | 8.84 | 0.30 | 0.62 | 2.03 | 1.62 | 0.37 | 0.27 | 1.16 | 0.42 |
| 453.6/1000 | 10.58 | 18.84 | 12.01 | 0.41 | 0.84 | 2.76 | 2.20 | 0.50 | 0.37 | 1.58 | 0.57 |

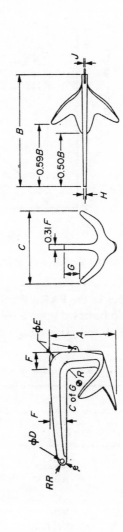

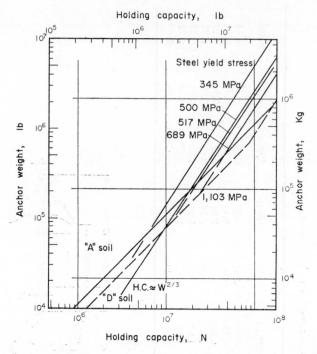

FIG. 6.   BRUCE anchor weights required to produce desired holding capacities.

advertised capacities of the PARAVANE or "Kite" anchor are provided in Table 4. These capacities appear to be based upon a soil with a shear strength approximately three times as strong as category A soil. This design does not lend itself to significant extrapolation due

TABLE 4.   CHARACTERISTICS AND COMPANY ADVERTISED HOLDING CAPACITIES OF THE PARAVANE ANCHOR

| | | Model No. | | | | |
|---|---|---|---|---|---|---|
| | | 200 | 500 | 1000 | 3000 | 6000 |
| | Weight kg | 94.8 | 225.4 | 449.1 | 1420.7 | 2717.0 |
| | (lb) | (209) | (497) | (990) | (3132) | (5990) |
| Size, m | h | 0.73 | 1.00 | 1.26 | 1.81 | 2.29 |
| | w | 0.98 | 1.34 | 1.69 | 2.44 | 3.09 |
| | l | 0.91 | 1.23 | 1.56 | 2.34 | 2.83 |
| | | Holding capacity, kN/kips | | | | |
| Embedment depth, m | 1.5 | 36/8 | 44/10 | 80/18 | 151/34 | 222/50 |
| | 3.1 | 53/12 | 93/21 | 138/31 | 302/68 | 445/100 |
| | 4.6 | 80/18 | 129/29 | 200/45 | 423/95 | 667/150 |
| | 6.1 | 98/22 | 178/40 | 285/64 | 583/131 | 898/202 |
| | 7.6 | — | 222/50 | 351/79 | 725/163 | 1125/253 |
| | 9.1 | — | — | 409/92 | 872/196 | 1352/304 |
| | 10.7 | — | — | — | 1023/230 | 1579/355 |
| | 12.2 | — | — | — | — | 1801/405 |
| | 13.7 | — | — | — | — | 2024/455 |
| | 15.2 | — | — | — | — | 2246/505 |

to the high bending stress created at the juncture of the plates. Rough estimates indicate that the maximum practical size would be in the 13.6 Mg (30,000 lbs) to 18 Mg (40,000 lbs) range with a capacity in category A soil of about 1.3 kN to 1.8 kN (0.3 to 0.4 × 10⁶ lbs).

Data on the DORIS Mud Anchor (C. G. Doris, 1973) are provided in Table 5. Advertised holding capacities are theoretically predicted; they are not based upon test data. In order to achieve these holding capacities with the shear strengths suggested by C. G. Doris (3.4 kPa to 9.6 kPa) the anchors must embed to depths approximately $2\frac{1}{2}$ times their height. Data would be needed to verify this penetration before confidence could be placed in the performance of the extrapolated anchors.

TABLE 5.   CHARACTERISTICS OF STANDARD DORIS MUD ANCHORS (DORIS, 1973)

| Type | Weight Mg/kips | Dimensions (m) | | | Advertised holding capacity kN/kips |
| | | Length | Width | Height | |
|---|---|---|---|---|---|
| A20 | 1.5/3.3 | 3.00 | 2.80 | 1.00 | 196/44 |
| A28 | 2.2/4.8 | 3.80 | 3.20 | 1.30 | 276/62 |
| A40 | 3.5/7.7 | 4.40 | 4.00 | 1.50 | 391/88 |
| A70 | 6.0/13.2 | 5.50 | 5.00 | 1.90 | 685/154 |
| A110 | 9.0/19.8 | 6.60 | 6.00 | 2.20 | 1076/242 |

The DORIS anchor appears to be suitable for very soft sediments where soil shear strength is relatively uniform or increases very slowly with depth. In normally consolidated soils like category A soil, there is a definite advantage to deep penetration. Bearing area for the DORIS anchor is roughly $1\frac{1}{2}$ times that for comparably weighted fluked and pick anchors, yet penetration of the latter would seem to be much greater. This would more than offset the area advantage of the DORIS. This is substantiated by the difference of more than a factor of 2 in efficiencies between the fluked or pick types and the DORIS anchor.

Large anchor-bearing areas are employed for the DORIS anchors to achieve high capacities with shallow burial. This results in low steel stress. The advantages of the DORIS anchor should be more fully realized in the very high load ranges where the fluked and pick type anchors become steel stress limited. The point at which this occurs in the DORIS anchor cannot be determined because plate thicknesses and steel types are unknown.

Extrapolation of the DORIS anchor to the ultra large sizes required by OTEC cannot be made with a great deal of confidence by maintaining geometric similarity. Extrapolation on this basis assumes that anchoring efficiency remains constant, until limited by steel stress. As dimensions increase by $W^{1/3}$, bearing area increases as $W^{2/3}$. Thus, for constant efficiency, penetration must increase. According to company data (Table 5), bearing area increases as $W^{6/7}$ to maintain constant efficiency (i.e. not geometrically similar). This means that fluke thickness is increasing at a rate less than $W^{1/3}$ which results in a more rapid increase in steel stress than would occur if similarity were maintained. In other words, the company probably assumed that penetration would not increase enough to maintain constant efficiency. Therefore, the increase in bearing area had to be disproportionately large (larger than $W^{2/3}$); and thickness relatively small. It is apparent that this offsets the advantage of maintaining low bearing pressure to produce low steel stress. Even though extrapolation is uncertain, the general size of some large anchors have been calculated and are presented in

Table 6 solely to enable estimates of fabrication and logistics difficulties. Assuming geometric similarity is not maintained, bearing areas and holding capacities were scaled according to $W^{6/7}$ and $23W^{0.9}$, respectively (Valent *et al.*, 1976). The ratio of fluke height to width was taken as 3/8.

TABLE 6.   CHARACTERISTICS OF EXTRAPOLATED DORIS MUD ANCHORS

| Anchor weight Mg/kips | Length (m) | Width (m) | Height (m) | Holding capacity kN/kips | Efficiency |
|---|---|---|---|---|---|
| 9.1/20 | 6.52 | 5.94 | 2.22 | 1080/240 | 12.1 |
| 18.1/40 | 8.81 | 7.99 | 2.99 | 1960/440 | 11.0 |
| 54.4/120 | 14.26 | 12.95 | 4.85 | 5250/1180 | 9.8 |
| 90.7/200 | 17.59 | 16.00 | 6.00 | 8450/1900 | 9.5 |
| 181.4/400 | 23.77 | 21.61 | 8.11 | 16,000/3590 | 9.0 |
| 453.6/1000 | 35.02 | 31.82 | 11.95 | 36,000/8100 | 8.1 |

The low efficiencies and large sizes of the DORIS compared to the fluke and pick type anchors previously discussed, seem to rule out this choice for OTEC applications. For example, assuming it could be fabricated, a 455 Mg (1,000,000 lb) mud anchor would be required to equal the capacity of a 136 Mg to 182 Mg (300 to 400,000 lbs) fluke type anchor, and the large size, at least 36 m (118 ft) long, 33 m (107 ft) wide and 12 m (40 ft) high, would make the mud anchor extremely difficult to handle and deploy.

*Anchor fabrication*

The limiting criterion regarding conventional anchor use in OTEC is fabrication capability. Two methods are available with which to fabricate these anchors, welding and casting.

The feasible construction limit for a welded anchor, without resorting to costly built up section, depends upon available steel. Heat treated steels of 760 MPa to 1100 MPa (110 to 160 ksi) strengths in thicknesses of 0.051 to 0.10 m (2 to 4 in.) can be obtained; however, the complexities involved with welding these materials make them undesirable choices. The most desirable steels for this application are HY-80 and A514. Of these, HY-80 is superior because of its superior toughness. Thicknesses of up to 0.11 m (4.5 in.) are available. This would allow a 68 Mg (150,000 lbs) fluke type anchor to be fabricated.

The feasible limit for a casting depends primarily upon the volume of steel required. Available information suggests that this would translate into approximately a 45 Mg to 64 Mg (100,000 to 140,000 lbs) maximum size for a pick or fluke type anchor. The practical limit for cast steel strength is 517 kPa (75 ksi). This value is almost coincident with the steel used to cast the Bruce anchor. One alternative is to fabricate an anchor by partial casting and welding. This would increase cost over a normal casting but could also potentially increase practicable size.

A significant factor in fabrication procedure is cost. For example, casting large anchors (45 Mg range) requires large flask capacity and would cost roughly $3.30 to $4.40 per kg including pattern fabrication. The cost of a welded construction of this size would be $5.50 to $6.60 per kg. The actual cost may be higher than the quoted figures because there would undoubtedly be some engineering involved in fabricating these large anchors.

The largest anchor that has been fabricated in this country is 27 Mg (60,000 lbs) (April, 1976) even though the largest advertised is 45 Mg (100,000 lbs).

Cost is not the only criterion that should be used in selecting fabrication method. Fabrication and use experience are also significant. There is limited experience in casting large anchors; thus, it appears that there would be more uncertainty both with fabrication and in-service use. Quality control of a very large casting may be difficult. Thus, the lower limit of the potential range (45 Mg) is recommended for casting.

*Summary*

Several standard drag-burial anchors including fluke, pick, and mud types were evaluated to determine the maximum practicable sizes that could be fabricated, handled, and deployed. The maximum size, 45 Mg to 68 Mg (100,000 to 150,000 lbs), will be controlled strictly by fabrication capability. The capability to handle and deploy components of this magnitude is well established.

Table 7 summarizes the characteristics of the possible anchors for OTEC. Maximum recommended sizes are 68 Mg (150,000 lbs) for a welded fluke type anchor and 45 Mg (100,000 lbs) for a cast anchor. Casting is the only reasonable method for fabricating the Bruce (pick type) anchor since its shape is not amenable to welded construction. A 64 Mg (140,000 lbs) cast anchor is conceded as a possibility within the present state-of-the-art with the realization that limited data exists in castings of this size. Anchor cost varies from 1 to 3 cents per newton (6 to 12c per pound) of holding capacity. In practice these differences would not be as significant because the higher cost is associated with a higher capacity anchor; thus, the number of mooring legs would be reduced with commensurate savings in materials and installation costs. Anchor size differences are not significant provided the fluke type anchor employs hinges in the stabilizer bars; otherwise, the approximately 12.2 m (40 ft) width would complicate deployment.

The drag burial anchors discussed are feasible to lateral capacities to 22 MN ($5 \times 10^6$ lbs). The difficulties inherent in using standard drag anchors in combination makes this type unsuitable for the Gulf Stream sites where loads to 178 MN ($40 \times 10^6$ lbs) are possible. It is very doubtful that eight of these large anchors could be placed in parallel such that each would be equally loaded.

Even in the deep water, benign environments where the lateral loads range only up to 18 MN ($4 \times 10^6$ lbs), drag anchors are not desirable. This is true because the drag embedment anchor requires a zero mooring line angle with the seafloor in order to embed, and requires that a near-zero angle be maintained in order to realize best holding capacity. Also, the drag embedment anchor is able to resist load from only one direction. These requirements of near-zero mooring line angles and uni-directional loading are too restrictive for general OTEC application.

The above requirements for drag embedment anchors can be satisfied. For instance, the uni-directional loading requirement can be met by using a multi-point moor in which each individual leg and its anchor experience service load is essentially from one direction. The requirement of zero mooring line angle during anchor embedment can be achieved in at least three ways:

1.  A sufficiently long scope of line could be used such that the mooring line angle at the seafloor was zero. However, the cost of the long line will be quite high, and the following techniques for shortening the scope will likely prove more cost effective.

TABLE 7. SUMMARY OF CHARACTERISTICS OF POSSIBLE DRAG EMBEDMENT ANCHORS FOR OTEC

| Anchor type | Fabrication method | Maximum weight suggested possible Mg/kips | Holding capacity MN/10⁶ lbs Cat. A | Cat. D | Holding capacity cost cents per N/cents per lb Cat. A | Cat. D | Size (m) Length | Width |
|---|---|---|---|---|---|---|---|---|
| Fluke | Welding | 68/150 | 16.9/3.8 | 22.2/5.0 | 2.7/12 | 2.0/9.0 | 10.67 | 3.96† / 11.58 |
|  |  | 45.4/100 | 11.1/2.5 | 15.6/3.5 | 1.8/8 | 1.4/6.0 | 9.14 | 3.35† / 9.75 |
| Fluke | Casting | 63.5*/140 | 15.1*/3.4 | 21.4*/4.8 | 1.8/8 | 1.4/6.0 | 10.36 | 3.96† / 11.28 |
| Pick | Casting | 45.4*/100 | 8.9*/2.0 | 12.0*/2.7 | 2.3/10 | 1.7/7.5 | 8.53 | 5.49 |
|  |  | 63.5*/140 | 13.3*/3.0 | 15.6*/3.5 | 2.3/10 | 1.8/8.0 | 9.75 | 6.40 |

*Maximum Casting Size, potentially involves quality control problems.
†Width with stabilizer hinged and folded.

2. Deadweight may be added to the mooring line near the anchor to "absorb" the vertical component of mooring line load. The deadweight may be mass blocks attached to the mooring line or may be a length of heavy chain incorporated into the lower end of the mooring line.

3. The drag embedment anchor could be fully embedded during installation using a long scope of line and zero mooring line angle, and then the scope shortened considerably raising the line angle to 0.1 rad (6°). Once embedded, the drag anchor will hold near capacity at this 0.1 rad line angle. This approach presumes that it is possible to preset an anchor to a few million pounds. Developing the necessary lateral force may be possible by pulling one anchor against another, but it will be difficult and time consuming. Further, if the present load was exceeded, the anchor could pull out.

Given that one of the above options for achieving a near zero line angle during embedment is invoked, an appropriate drag embedment anchor could be built. An anchor such as the Bruce would be desirable because it would reorient itself and embed regardless of initial attitude. Maximum lateral capacity would then be limited to 91–6 MN (2–3.5 × 10$^6$ lbs) without resorting to multiple anchor hookups.

## Conclusions

Of the two anchor types considered potentially applicable to OTEC, the fluke and pick types, the least complicated installation would be achieved with the pick type because of its ability to embed irrespective of landing attitude. The maximum capacity of this anchor would be 9 to 16 MN (2 to 3.5 million pounds). If the more complicated installation procedures associated with the fluke type are acceptable, then 13 to 22 MN (3 to 5 million pounds) capacity is possible. For these anchors to be useful, they would have to be used in multi-point moorings possibly with several anchors in tandem or parallel in each leg.

Until fabrication technology improves to a point where 114 Mg to 182 Mg (250,000 to 400,000 lbs) anchors are feasible, the use of conventional anchors of the pick and fluke types would seem to be less attractive than other types of anchors for the OTEC installation.

## PLATE ANCHORS

## Introduction

Several plate anchors were designed and analyzed for holding capacity. Calculations were made for seafloor soil categories A, C, and D. Possible OTEC plate anchor designs were also compared to existing plate anchor technology. Rough cost estimates were made on the basis of present material and fabrication costs.

## Holding capacity

Soil categories A and C (cohesive soils) were analyzed for both long term and short term behavior under static loading. Soil category D (cohesionless) was analyzed for long term static loadings. Anchors considered were flat plates with areas from 9.3 m$^2$ to 149 m$^2$ (100 ft$^2$ to 1600 ft$^2$). Calculations were made for both square and rectangular (length = 2 × width) plates. Holding capacities at anchor embedment depths of 6.1 m to 30.5 m (20 ft to 100 ft) were determined. The procedure proposed by Taylor and Lee (1972) was used.

*Short term holding capacity.* Short term static loading describes the situation in which the anchor is loaded rapidly until breakout occurs. For short term loadings the holding capacity to be achieved is calculated differently for cohesive soils (clays) than for cohesion-less soils (sands).

In cohesive soils the excess soil pore pressure generated by rapid loading requires a lengthy time period for near-complete dissipation. Because the pore pressure beneath an embedment anchor plate does not have sufficient time to dissipate during rapid loading, suction develops under the plate causing an apparent increase in holding capacity. The magnitude of this suction force cannot be predicted confidently for the highly disturbed soil expected about the large OTEC plate anchors. Because of this lack of confidence, the contribution of the suction force to short-term holding capacity was herein conservatively assumed to be zero; i.e. no suction.

In cohesionless soils the excess soil pore pressure due to rapid loading is dissipated almost as fast as it is generated. Thus, a suction force is not generated.

The basic equation used for holding capacity is a typical bearing capacity equation (Taylor *et al.*, 1975):

$$R = A(s\overline{N}_c + \gamma d\overline{N}_q)(0.84 + 0.16 \, B/Z), \qquad (2)$$

where $R$ = holding capacity (N)
$A$ = plate area (m²)
$s$ = soil cohesion (Pa)
$\gamma$ = buoyant unit weight of soil (N/m³)
$d$ = plate embedment (m)
$\overline{N}_c, \overline{N}_q$ = holding capacity factors
$B$ = fluke width (m)
$Z$ = fluke length (m)

For cohesionless soils, $s = 0$ and $\overline{N}_c = 0$. For cohesive soils under short term loading, $\overline{N}_q = 1$ and $\overline{N}_c$ ranges from a value of zero to nine.

*Long term holding capacity.* Long term static holding capacity refers to the situation in which an anchor pulls out after a constant upward force has been applied over a long period of time. Such a situation might occur for a submerged buoy moored to the seafloor (Taylor *et al.*, 1975). Equation (2) with $s$ and $\overline{N}_c$ set to zero was used to find long term capacity. $\overline{N}_q$ ranged from two to twelve. Under long term loading, the excess pore pressure in cohesive material escapes (drained behavior). As a result, clays under long term loading act as frictional materials. Cohesionless materials were assumed to exhibit drained behavior only.

*Design holding capacity.* Short and long term holding capacities were plotted against depth and plate size in Figs 7, 8 and 9. For cohesive soils the short term capacity was much less than predicted long term capacities. Therefore, short term capacity was taken as the critical or design holding capacity. The curves show the large plate sizes and deep embedment depths required to achieve OTEC holding capacities. The single plate size required in the deep ocean environment, soft clay (category A), is 12.2 m wide. Compare this with the largest existing embedment anchor plate which measures 1.5 m wide to 2.5 m long.

*Effect of inclined load.* Another consideration that enters into a single point mooring (single or multiple flukes) is the effect of inclined or non-vertical loads. Several studies

(Meyerhof, 1973; Colp and Herbich, 1972) on the effects of inclined loads have been conducted. For almost all soil types and relative depths of embedment Meyerhof found that holding capacity under an inclined load equalled or exceeded that under a vertical load. Tests by Colp and Herbich in a saturated sand showed that capacity increased up

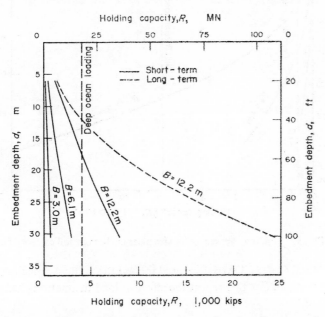

FIG. 7.   Holding capacity of square plate anchors in soil category A.

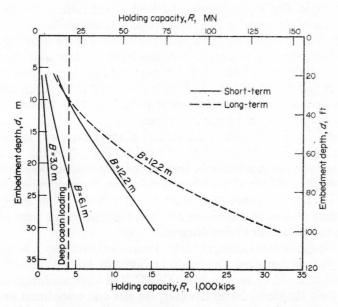

FIG. 8.   Holding capacity of square plate anchors in soil category C.

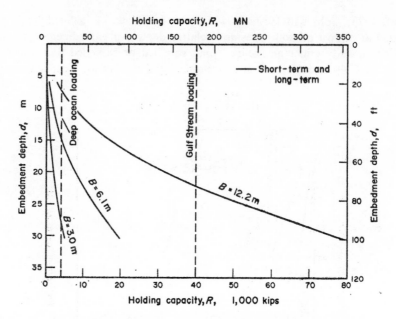

FIG. 9.   Holding capacity of square plate anchors in soil category D.

to an angle of 0.44 rad (25°) and then decreased to near the vertical capacity at an inclination of 0.79 rad (45°) vertical. For this report the effect of load inclination changes was assumed to be negligible.

### Operational factors

*Mooring line angle.* For a given horizontal load at the surface, the load felt at the anchor is a function of the mooring line angle (angle between seafloor and mooring line). For a taut line the load at the anchor increases approximately as the secant of the line angle at the seafloor. Thus, for high line angles mooring loads are increased significantly. For example, a horizontal force of 18 MN ($4 \times 10^6$ lbs) would produce a force in the mooring line about 25 MN ($5.7 \times 10^6$ lbs) at a line angle of $\pi/4$ rad (45°). At an angle of 1.4 rad (80°) the force in the mooring line would be roughly 102 MN ($23 \times 10^6$ lbs). The advantage of maintaining a low mooring line angle is apparent.

*Keying Distance.* Keying distance is the vertical distance (upward) required to rotate the anchor to an approximately horizontal orientation (Fig. 10). A rule of thumb is that the keying distance equals twice the fluke length. For example, suppose that a 6 m wide fluke must be embedded to 20 m to mobilize soil resistance. To account for keying distance, the fluke length, i.e., required penetration = 20 m + 2 × 6 m = 32 m. For typical OTEC relative embedments ($d/Z = 5$) this distance becomes excessive. The large plates must be driven very deep to account for keying distance.

*Single Plate.* Design anchor loadings (Table 8) were used with Figs 7, 8, and 9 to specify required OTEC plate anchor sizes. The results are listed in Table 9. Table 9 suggests that a single plate anchor could be used to moor OTEC. The basic advantage of these single fluke designs is their simplicity. Note, however, the size and embedment depth required. These sizes represent an enormous extrapolation of current plate anchor installation

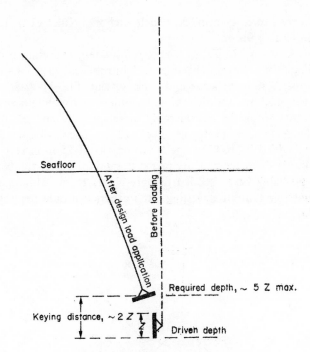

FIG. 10.   Illustration of keying distance.

TABLE 8.   OTEC ANCHOR LOADING

|  | Deep ocean | Gulf Stream |
|---|---|---|
| Possible soil category | A, B, C | D, E, F |
| Assumed maximum mooring line angle with seafloor | 0.79 rad (45°) | |
| Maximum horizontal force | 18 MN (4 × 10⁶ lbs) | 180 MN (40 × 10⁶ lbs) |

TABLE 9.   PLATE ANCHOR DESIGN FOR OTEC APPLICATION

| Load MN/10⁶ lbs | Soil | Width* (m) | Thickness (mm) | Required depth (m) | Driven depth (m) | Number† required |
|---|---|---|---|---|---|---|
| 25.5/5.6 | A | 3.05 | 83 | 30 | 37 | 9 |
|  |  | 6.10 | 142 | 30 | 43 | 2 |
|  |  | 12.19 | 203 | 24 | 48 | 1 |
| 25.5/5.6 | C | 3.05 | 142 | 30 | 37 | 4 |
|  |  | 6.10 | 203 | 30 | 43 | 1 |
|  |  | 12.19 | 203 | 14 | 38 | 1 |
| 255/56 | D | 3.05 | 229 | 30 | 37 | 12 |
|  |  | 6.10 | 457 | 30 | 43 | 3 |
|  |  | 12.19 | 668 | 26 | 50 | 1 |

*Square in Plan.
†Factors of Safety not included.

technology. The means to embed and key such anchors neither exists today nor appears possible in the foreseeable future.

*Multiple Plate.* As shown in Table 9, required holding capacities for OTEC could be obtained by bridling several flukes. This approach introduces the problem of load equalization among anchors. A simple sheave and cable system (Fig. 11) has been used to solve this problem for two anchors. But designing, installing, and maintaining a bridling system for up to six anchors would be considerably more intricate and costly. Note also that the driven depth for 3 m flukes in category A soil is 37 m. The Civil Engineering Laboratory has explosively embedded its 100 kip anchor to depths of 15 m in soft seafloors using a 254 mm (10 in.) inside diameter gun. By increasing gun diameter to 406 mm (16 in.), the 3 m plate could probably be embedded to only about 15 m. Although other means of embedment are possible (vibrating, jetting, etc.) experience indicates that they would be even less desirable.

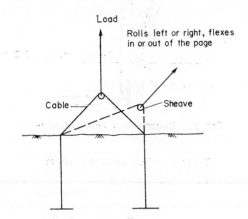

FIG. 11.   Load equalization for two-anchor case.

The load equalization and embedment problems associated with multiple plate anchor installations would require a concentrated and expensive design and development program with no guarantee of success. Such a program is not recommended in view of the alternative anchor types available.

*Structural design*

*Method.* The simple model shown in Fig. 12 was used to compute stresses in plate anchors. Mooring force was assumed to act vertically. It was balanced by a uniform distribution of soil pressure on the horizontal plate. Maximum bending moments and resulting fiber stresses were calculated and used to compute required steel thickness. Under the assumptions, required plate thickness is independent of plate area. A plot of mooring force vs required plate thickness (HY-80) steel is shown in Fig. 13. From the plot it is clear that steel of 76 mm to 101 mm (3 to 4in.) thickness is required.

To reduce the steel weight (cost) imposed by large plate thicknesses, a hollow plate design was also considered. Typical solid and hollow designs are shown in Fig. 14.

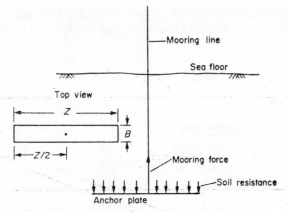

FIG. 12.  Model for plate anchor structural design.

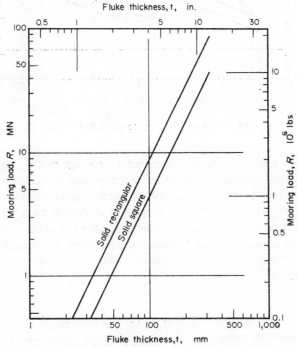

FIG. 13.  Required fluke thickness vs mooring load.

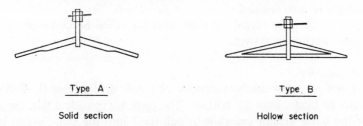

FIG. 14.  Cross-section of solid and hollow plate anchor design.

Anchor loading was plotted against anchor weight for both solid and hollow designs in Fig. 15. The reader should note that this figure does not indicate anchor holding capacity.

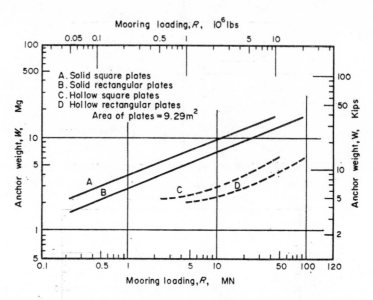

Fig. 15.  Required anchor weight vs mooring load for solid and hollow anchors.

It merely gives the weight (thickness) required to prevent steel failure under an applied load. Anchor capacity will be a function of soil type and embedment depth as shown in Figs 7, 8, and 9. The nine square metre (100 ft²) anchor has maximum holding capacity of 3.3—N (0.75 × 10⁶ lbs). Now Fig. 15 is entered to determine anchor weight.

*Cost*. Table 9 shows several possible anchor designs for OTEC applications. The 3 m square anchor was chosen for cost comparison to present anchors. The estimated cost of fabrication and materials only for a deep ocean plate anchor installation (nine 3 m anchors) is $234,000 (Table 10).

### Summary

Plate anchors, either singly or in arrays, are not reasonable choices for OTEC. Load equalization for multiple fluke designs and embedment limits for either multiple or single fluke design are critical. In view of the alternative anchors available further investigation of plate anchors is not justified.

Existing or moderately scaled up plate anchors could be useful in conjunction with another primary anchoring system.

### Screw-in plate anchors

Present types of screw anchors consist of small diameter steel shaft with a large diameter plate, or blade, near the bottom. The plate is typically a flat, single-flight helix. Rotation of the screw anchor causes it to pull itself into the soil. Several blades may be spaced along the shaft for greater capacity. Screw anchors have been used extensively on

TABLE 10.   MATERIALS AND FABRICATION COST ESTIMATE FOR A
TYPICAL OTEC PLATE ANCHOR

| | |
|---|---|
| **Environment—Deep Ocean** | |
| Soil category | A (soft clay) |
| Water depth | 6000 m |
| Total mooring force | 18 MN (4 × 10⁶ lbs) |
| | |
| **Anchor Description—nine 3 m square plate anchors** | |
| Nominal holding capacity (each)—3 MN (0.67 × 10⁶ lbs) | |
| Dimension | 3 m × 3 m × 83 mm |
| Weight | 5900 kg (13,000 lbs) |
| Efficiency | 51 |
| | |
| **Cost** | |
| Hy—80 steel | $0.77/kg |
| Fabrication (Welding) | $3.63/kg |
| Unit Cost | $4.40/kg |
| Sub-total (1 anchor) | $26,000 |
| Total (9 anchors required) | $234,000 |

land as guy anchors. They have also been used to anchor offshore pipelines. Present offshore systems are installed in water depths to 300 m and have a pullout capacity of about 110 kN (25 × 10³ lbs).

Relatively minor modifications are required to increase capacity by a factor of 2 (Raecke, 1973). However, scaling up to the sizes required for OTEC would be a monumental task. Used as the primary anchor, a screw-in anchor would be limited by lateral load capacity. As discussed in the section on piles, the sizes required to resist lateral loads are probably unreasonable. For this reason screw-in anchors will be considered only as supplementary anchors to assist in resisting vertical load.

*Proposed use.* Single helix anchors might be used to supplement the vertical holding capacity and resistance to overturning of a deadweight anchor (Fig. 16). The screw anchor and associated embedment system would be fabricated as an integral part of the deadweight. The deadweight and screw anchor combination would then be transported and positioned as a single unit. The helix would be embedded by hydraulic motors driven by electric power from the surface as proposed by Raecke (1973).

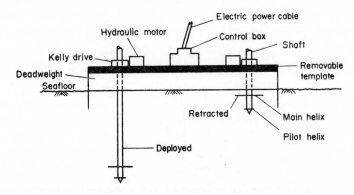

FIG. 16.   Composite deadweight and screw-pile anchor, resists high uplift loads.

Only a single helix arrangement is considered feasible for the procedure suggested above. The added capacity of a multiple helix is outweighed by the complexities of installing it through the deadweight.

*Holding capacity.* The holding capacity of a single helix anchor can be computed using procedures identical to the plate anchor procedures presented above. Approximate helix areas required may be found using Figs 7, 8, and 9. The representative plate anchor sizes in Table 9 may also be taken as representative helix sizes.

*Embedment.* The power required to embed helices of the sizes suggested by Table 9 is a matter of conjecture. The screw-in anchor concept proposed by Raecke required a 45 kW (60 hp) power source at the surface to install a single-helix anchor, 0.8 m diameter, to a depth of 15 m. Present offshore systems (Short, 1971) are powered by hydraulic motors requiring about 56 kW (75 hp). On the basis of fluke area, or holding capacity, an extrapolation of roughly 20 : 1 would be required for OTEC helix sizes. Power requirements are probably within reach.

*Summary*

1. Single helix anchors, used to supplement the vertical capacity of a deadweight, are feasible. Further investigation should provide information regarding the economic aspects of the suggested concept.

2. Single or multiple helix anchors are not feasible as the primary OTEC anchor due to low lateral load-capacity.

3. Multiple-helix anchors do not appear suitable for use with deadweight because of installation complexities.

## PILE ANCHORS

The evaluation of piles as OTEC anchors first considered a single pile, unrestrained head, subjected to a slowly applied horizontal and vertical load (Fig. 17). Lateral and axial (pull-

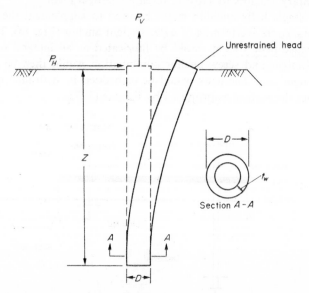

FIG. 17. Deflection of a long pile by combined axial and lateral load applied at seafloor surface.

out) capacities were determined for soil categories A, B, C, and D. Behavior of piles subjected to repetitive loading was estimated using available guidelines.

*Pile lateral capacity*
    The effect of soil strength, pile length, and pile diameter on lateral capacity was deter-
mined by combining the results of a short and a long pile analysis. Overlap of the two
procedures at a transition length (10 × diameter) allowed some comparison of results.
Comparisons for piles of a given diameter were made for similar pile stiffnesses.
    *Short pile analysis.* Czerniak's (1957) procedure for analyzing short, rigid piles was used.
The procedure is based on the principle of static equilibrium. The pile is assumed to be
infinitely rigid. It rotates about a point below the soil surface when an external lateral force
is applied at the top of the pile (Fig. 18). Soil resistance at each point along the pile is
assumed to be proportional to pile deflection at that point.

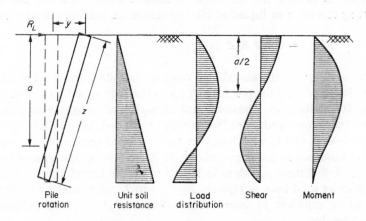

FIG. 18.   Short pile loading and displacement.

$$p = Ky/D \tag{3}$$

where $p$ = soil pressure (Pa)
    $K$ = modulus of horizontal subgrade reaction (Pa)
    $y$ = lateral deflection (m)
    $D$ = outside diameter of the pile (m).

Assuming a linear increase of unit soil resistance with depth, the length of pile necessary
to maintain equilibrium under an applied load can be calculated.
    For a horizontal load $R_L$, and moment $Mo$, applied at the surface, the appropriate
equation of equilibrium is:

$$Z^3 - 14.14 \frac{R_L Z}{Dn_h} - 18.85 \frac{Mo}{Dn_h} = 0 \tag{4}$$

where $Z$ = pile length (m)
    $R_L$ = lateral force at the surface (N)
    $n_h$ = coefficient of horizontal subgrade modulus (Pa/m)
    $Mo$ = moment at the surface (Nm)

For the special case of no external moment, Equation (4) becomes:

$$R_L = \frac{Z^2 D n_h}{14.14} .$$

(5)

Similarly, the equation for bending moment at any point is:

$$M_x = R_L Z \left\{ \left(\frac{x}{Z}\right) - 3 \left(\frac{x}{Z}\right)^3 + 2 \left(\frac{x}{Z}\right)^4 \right\}$$

(6)

where   $x$ = distance along pile below the surface (m)
   $M_x$ = moment at point $x$ (Nm).

For $R_L$ applied at the surface, the maximum bending moment occurs at approximately 0.42$Z$. Substituting this for $x$ in Equation (6), the maximum bending moment becomes:

$$M_{max} = 0.260 \, R_L Z.$$

(7)

Equation (5) was used to find the lateral load capacity for pile diameters for 1.2, 2.4, 4.9 and 7.6 m (4, 8, 16, and 25 ft) for pile lengths up to the limit of validity of the Czerniak relationship, i.e. 10 × diameter. The coefficient of horizontal subgrade modulus, $n_h$, was selected based on the recommendations in Czerniak (1957) and Anon (1967). Those values for Soil Categories A, B, and C were based on the developed shear strength profiles, and the value for Soil Category D based exclusively on Anon (1976). The assumed values are listed in Table 11. Using these $n_h$ values in Equation (5) the lateral load vs pile strength relationships shown as solid lines in Figs 19, 20, 21, and 22 were developed. Soil pressures were assumed to be sufficient to maintain equilibrium and resulting deflections were assumed to be acceptable.

TABLE 11. ASSUMED COEFFICIENTS OF
HORIZONTAL SUBGRADE MODULUS

| Soil category | $n_h$ (kPa/m) | (psf/ft) |
|---|---|---|
| A (Clay) | 14.5 | 92 |
| B (Clay) | 21.8 | 139 |
| C (Clay) | 30.8 | 196 |
| D (Sand) | 84.8 | 540 |

*Long pile analysis.* A long pile analysis was done for two reasons:
(1) To find out if pile capacity could be increased using longer piles.
(2) To check the short pile results at the transition length ($Z = 10D$).

A subgrade reaction method (Gill and Demars, 1970) was used to find lateral capacity for selected pile lengths and diameters. The method is similar to a commonly used method developed by Matlock and Reese (1960). The non-linear soil load–deformation characteristics ($p$–$y$ curves) are represented by a rectangular hyperbola. Hyperbola shape is defined by measurement of the undrained shear strength for clay; and effective overburden

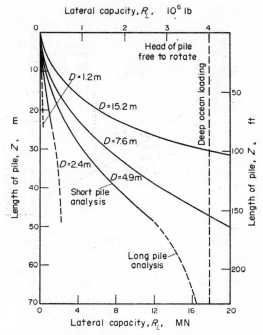

FIG. 19.  Pile lateral capacity in soil category A.

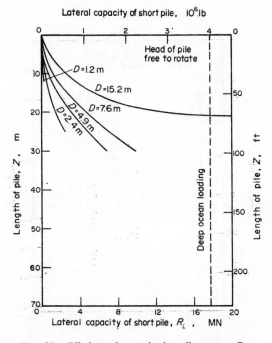

FIG. 20.  Pile lateral capacity in soil category B.

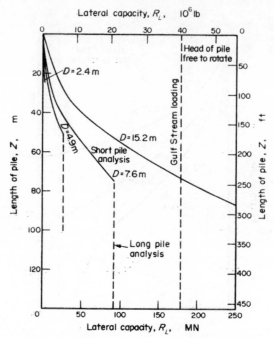

FIG. 21.   Pile lateral capacity in soil category C.

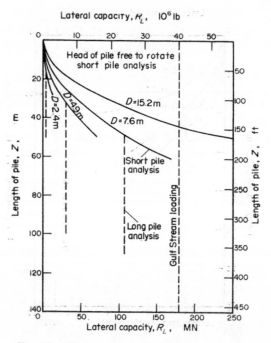

FIG. 22.   Pile lateral capacity in soil category D.

and friction angle for sand. The procedure employs a finite difference solution of the equation of bending to obtain deflections. Soil properties, pile characteristics, and pile loading are input to the computer program. Pile stiffnesses were selected on the basis of maximum bending moments calculated from the Czerniak rigid-pile analysis for soil category C. The pile sections required to obtain these stiffnesses appear reasonable and practical for fabrication. The long-pile analysis program returns pile deflection, bending moment and soil modulus as a function of depth. The external loading boundary condition was adjusted until pile failure (bending moment exceeded) occurred.

Long pile analysis lateral load capacities for various lengths and diameters are shown as dashed curves in Figs 19, 21, and 22. A long pile analysis was not done for category B soil (Fig. 21). Curves were drawn through the long and short pile analysis, i.e. at $Z = 10 \times D$. The solid and dashed lines delineate the lateral capacity envelope for the piles shown.

*Results.* Review of Figs 19 and 20, indicative of pile anchor performance in deep ocean environments, suggests that single piles of 7.6 m diameter (25 ft) and 47 m in length would be required to mobilize the necessary 18 MN ($4 \times 10^6$ lbs) lateral load resistance.

Figures 21 and 22, indicative of performance in a competent clay and sand seafloor, as might be found beneath a high energy area, suggests that a single pile anchor design is probably not appropriate for such an area. Piles of 7.6 m diameter (and of the stiffnesses assumed) are not capable of resisting the full 180 MN horizontal load component ($40 \times 10^6$ lbs). Larger diameter piles could mobilize sufficient soil resistance to resist the Gulf Stream loading; however, technical feasibility suggests that a multiple pile anchorage using smaller piles would be more reasonable.

Figure 23 illustrates the influence of the pile analysis procedure employed. The short pile analysis assumed mobilization of the available soil resistance and maximum bending moment in the short pile was calculated. Then a pile section was designed to resist that

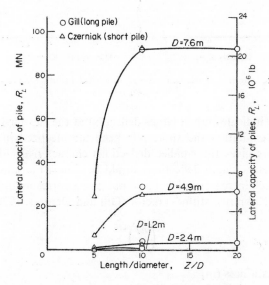

Fig. 23.  Pile lateral capacity vs normalized length.

moment. This pile section was then input into the long pile analysis (or more properly the stiffness values were input). This fixing of the pile stiffnesses results in the fixing of the bending resistance of the respective pile diameters. For lengths greater than about 10 diameters, lateral load capacity is controlled by the assumed bending resistance. Adherence to this procedure is justified because the pile sections designed appear practical as is: increasing the pile stiffnesses would result in very thick-walled, unreasonable, possibly unfabricatable sections. The actual pile sections designed will be presented and discussed later; it is important to note here that the break in the lateral load capacity curve (Fig. 23) is not indicative of a soil phenomenon, but rather is due to the pile stiffness assumed in the analysis. The data presentation of Fig. 23 suggests that for the pile sections used, there is no marked improvement in lateral load capacity when the pile is lengthened beyond 10 diameters. However, increasing the length beyond 10 diameters may decrease pile deflection. Figure 24 shows deflection at maximum load for the 8 ft diameter pile at $Z = 10D$ and

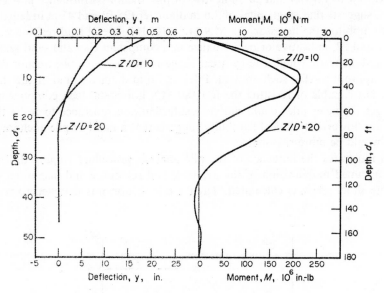

Fig. 24. Pile deflection and moment vs soil depth (at failure loading) for the 2.4 m diameter pile.

$Z = 20D$. The figure indicates much larger deflection at the shorter length. At $Z = 10D$, the pile is definitely exhibiting the short pile behavior idealized in Fig. 18. Increasing length to 20 diameters gives the smaller deflections characteristic of long pile behavior (Fig. 17). At 30 diameters length deflections would be almost the same as the 20 diameter case. The point at which increases in length do not give a corresponding decrease in deflections depends on a well-known stiffness factor (Gill and Demars, 1970, and others):

$$T = \left(\frac{EI}{n_h}\right)^{1/5} \tag{8}$$

where   $T$ = relative stiffness (m)
$\quad\quad\;\; E$ = Young's modulus for the pile (Pa)

$I$ = polar moment of inertia of the pile (m⁴)

$n_h$ = coefficient of horizontal subgrade modulus (Pa/m).

Piles with embedded lengths larger than 4T act as if they were embedded to an infinite depth (Davisson and Gill, 1963).

The desirability of small deflection takes on increased significance when repetitive loading is present. This will be discussed more in the section on repetitive loading.

### Pile axial capacity

*Pullout capacity in clay.* A semi-empirical procedure (Vijayvergiya and Focht, 1972) was used to predict the pullout capacity of the piles for cohesive soils (A, B, C). The method of analysis expresses pile frictional resistance as a function of mean vertical effective stress and mean undrained shear strength:

$$Q_s = \lambda \, (\bar{\sigma}_m + 2s_m) \, A_s \qquad (9)$$

where  $Q_s$ = side friction on pile (N)

$\lambda$ = dimensionless friction coefficient

$\bar{\sigma}_m$ = mean vertical effective stress over pile length (Pa)

$A_s$ = lateral area of embedded pile (m²)

$s_m$ = mean undrained shear strength over length of pile (Pa).

The friction coefficient was based on available data from load tests at a number of different locations. Comparison of predicted friction pile resistance to observations showed good agreement (Vijayvergiya and Focht, 1972).

*Pullout capacity in sand.* For category D soil (sands) a simplified empirical procedure was followed. Pullout resistance was taken as a function of mean vertical effective stress, $\bar{\sigma}_m$:

$$Q_s = K_s A_s \, \bar{\sigma}_m \tan \delta \qquad (10)$$

where $K_s$ = coefficient of lateral pressure on pile wall

$\delta$ = effective friction angle of soil on pile wall (rad).

For the calcareous sand, $\delta$ was taken as 0.44 rad (25°) and $K_s$ as 0.4. Such high values are appropriate for calcareous materials only when the piles are installed by drilling and grouting (McClelland, 1974). Available large scale test information (Angemeer *et al.*, 1973, 1975) indicates that capacities of driven piles in calcareous sands may be as little as one-fourth of those indicated by Equation 10. (Pile load tests are required to accurately predict the capacity of driven piles on this type material.) The size and difficulty of installation of a pile anchor for the OTEC power plant dictates that the soil resistance available be maximized. Therefore, this study assumes that the technology will be developed to provide the drilled and grouted pile option in water depths to 6,000 m. In reality, the installation of large anchor piles in deep water appears more technically feasible by the drilling and grouting technique than by the driving technique.

*Results.* Computed pullout capacities are shown in Figs 25, 26, 27, and 28 for respective soil categories. The reader should note that capacities were calculated independently of lateral load. Again, calculations were for a single, unrestrained pile subjected to a slowly applied vertical load. Note that a 110 m long, 4.9 m diameter pile would provide sufficient pullout resistance in the deep ocean environment. An 85 m long, 7.6 m diameter

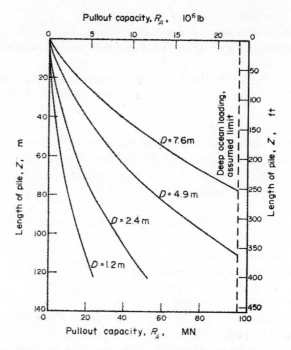

FIG. 25.  Pile axial pullout capacity in soil category A.

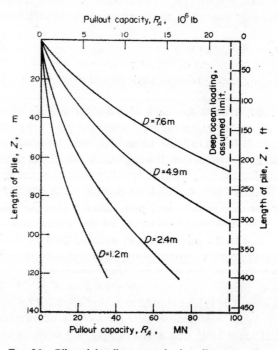

FIG. 26.  Pile axial pullout capacity in soil category B.

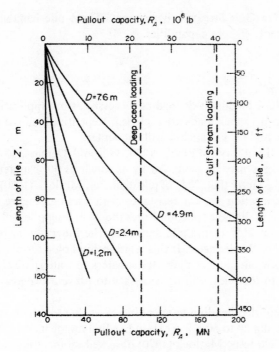

FIG. 27.   Pile axial pullout capacity in soil category C.

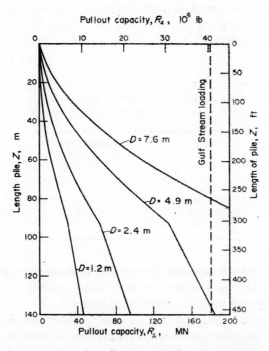

FIG. 28.   Pile axial pullout capacity in soil category D.

pile would hold in the Gulf Stream environment. These pile lengths and diameters are probably within present offshore capabilities.

### Effect of repetitive loading

*Definition.* Repetitive loading is defined as successive, slow applications and relaxations of a load. In other words, shock loads are excluded. Repetitive load is a reasonable approximation of what an actual OTEC anchor will encounter.

*Discussion.* Several authors (Matlock, 1970; Gill and Demars, 1970) have noted that static load displacements may be increased by as much as 50% under repetitive loading. At initial displacements larger than 20% of pile diameter, Matlock (1970) noted a continuous and progressive deterioration of soil resistance with cycling. Figure 24, presenting displacement data for 2.4 m diameter pile, indicates the 24 m long pile will deflect 0.5 m under static loading, or about 21% of the pile diameter. The longer 48 m pile will deflect only 0.26 m or about 11% of the diameter. If this relationship holds for piles of other diameters and in other soils, then piles longer than 10 diameters (or, alternatively, reductions in the lateral loads applied to the piles) will be necessary to prevent progressive deterioration of soil resistance.

Several authors reported bending moment increases due to repetitive loading. Davisson and Salley (1968) noted a 20 to 50% increase in bending moments on 1.2 m diameter drilled piers. In model and field tests, Matlock (1970) observed moment increases of almost 100% during cyclic loading. Several of Matlock's tests were performed on a 0.33 m diameter instrumented pile in soft, slightly over-consolidated clay. These conditions are fairly representative of marine sediment deposits. Matlock suggested that a reduction in pile lateral capacity due to cycling might be estimated on the basis of a ratio of pile deflection to diameter. The rough guidelines in Table 12 were estimated from the results of his model tests.

TABLE 12. SUGGESTED RESIDUAL LATERAL LOAD CAPACITY PERCENTAGE DUE TO REPETITIVE LOADING AS A FUNCTION OF FIRST CYCLE DEFLECTION

| $y/D$ | Lateral capacity (per cent of static capacity) |
|-------|-----------------------------------------------|
| 0.2   | 20 |
| 0.1   | 58 |
| 0.05  | 62 |
| 0.025 | 67 |

A typical deflection ratio expected for the OTEC anchor is about 0.1. Thus, design capacity for repetitive loading is about 0.58 times static capacity. Please note these reductions have not been applied to the data presented herein.

The effect of repetitive axial loading on axial pullout capacity should be minimal (Bea, 1975). Likewise the effect of repetitive lateral load on axial capacity has not been a serious problem in present offshore structures.

*Preliminary structural design of required piles*

As indicated in the lateral load capacity analysis section, the pile sections were designed to resist the maximum load conditions for piles with lengths equal to 10 diameters in the category C soil. This course of action was taken early in the study to enable an evaluation of the materials available for fabrication, the wall thicknesses and weights necessary, and the overall practicality of the pile anchor concept applied to OTEC. Table 13 presents a

TABLE 13.   PILE ANCHOR CHARACTERISTICS, CATEGORY C SOIL

| Diameter, m | 4.9 | 4.9 | 7.6 |
|---|---|---|---|
| Length, m | 31 | 31 | 76 |
| Environment of possible use | Deep Ocean | Deep Ocean | Gulf Stream |
| Axial load, MN/ $\times$ 10⁶ lbs | 22.7/5.1 | 22.7/5.1 | 149/33.6 |
| Lateral load, MN/ $\times$ 10⁶ lbs | 9.8/2.2 | 9.8/2.2 | 98/22.0 |
| Material | Prestressed concrete | Steel | Steel |
| Grade | $f_c' = 41$ MPa (6000 psi) | A-36 | HY-80 |
| Cross section | Solid | Shell | Shell |
| Wall thickness, mm | — | 51 | 152 |
| Maximum stress, MPa/ksi | — | 104/15 | 321/47 |
| Number of 42.8 mm diameter strands | 188 | — | — |
| Total weight, Mg/ $\times$ 10⁶ lbs | 1400/3.0 | 185/0.41 | 2100/4.7 |
| Lateral load/weight | 0.7 | 5.4 | 4.6 |

summary of results for two, potential, single-pile anchor designs, one a 4.9 m diameter, 31 m long pile, the other a 7.6 m diameter, 76 m long pile. These piles in a category C soil are capable of mobilizing soil resistance to supply the noted axial and lateral load capacities, with no reductions applied for repetitive loading. In all cases, the large bending moments resulting from the high lateral loads controlled structural design of the piles.

Reinforced concrete was eliminated from consideration because the high bending moment combined with axial tension would result in significant cracking on the tension side of the pile, an unacceptable condition in the marine environment. Either prestressed concrete or steel could be used for the 4.9 m diameter pile resisting the 9.8 MN (2.2 $\times$ 10⁶ lbs) lateral load. The prestressed concrete pile would, however, require 13.3 MN (3 $\times$ 10⁶ lbs) of steel and concrete to resist the 9.8 MN lateral load. Thus, the concrete pile anchor efficiency (lateral load capacity to weight ratio) would be 0.7—little better than a good deadweight anchor. The steel pile appears much more desirable with a 50 mm wall thickness and a lateral load efficiency of 5.4.

An adequate design using prestressed concrete is not possible for the 7.6 m diameter pile resisting the 97 MN (22 $\times$ 10⁶ lbs) lateral load. One simply cannot fit the required steel and concrete into the cross-section available. An idealized steel pile section is suggested for illustrative purposes. The 7.6 m diameter pile would require pile walls 150 mm (6 in.) thick. HY-80 steel can be obtained in such thickness, is weldable, and would serve well in seawater at the ambient 4°C; it is not known if such plates can be formed to the necessary shell curvature. Of course, the real pile structure would probably look considerably different from the idealized shell. One fact is known: the pile structure will be steel and will not be reinforced or prestressed concrete.

*Summary*

Probable anchor pile ultimate capacities are shown in Fig. 29 for the deep ocean environment and in Fig. 30 for the Gulf Stream environment. In reviewing and using this ultimate capacity data, the reader must remember that safety factors on the order of 2 must be

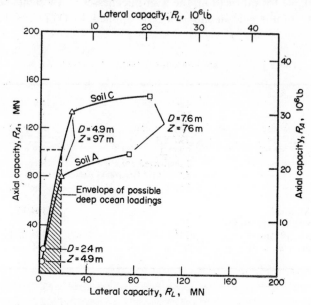

Fig. 29.   Pile anchor holding capacity under static load, deep ocean environment.

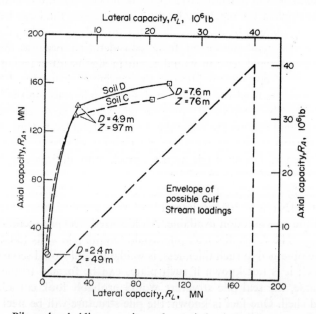

Fig. 30.   Pile anchor holding capacity under static load, Gulf Stream environment.

applied to this data when determining the safe working load. Working loadings higher than this will cause excessive pile deflections leading to pullout or failure in bending. Capacities for axial and lateral load were computed independently. Pile lengths plotted in Fig. 29 were chosen to: (1) minimize deflection and (2) reflect practical constraints (installation, fabrication, anticipated costs). Figure 29 suggests that single piles used to anchor OTEC power plants in the deep ocean will have to be about 4.9 m in diameter to resist the lateral load component. The pile length required would vary with the vertical load component, from about 50 m long for zero vertical load to about 120 m long for a 100 MN vertical load component (see Fig. 25 for source of pile length data).

Figure 30 indicates the difficulty of obtaining the very high lateral load capacities necessary for the Gulf Stream environment.

Pile anchors of conventional diameter, say 2.4 m, linked in groups do not appear economical for resisting the large lateral forces encountered in locations like the Gulf Stream. To resist repetitive lateral load with piles alone would require as many as 118 2.4 m diameter, 49 m long piles.

In the less severe deep ocean environment piles could prove to be effective and economical. They have one distinct advantage over deadweights. Pile lateral capacity is not severely reduced as axial pullout forces increase. Thus piles are better able to cope with high mooring line angles than are deadweight or drag embedment anchor systems.


## DEADWEIGHT ANCHORS

*Introduction*

*Horizontal load resistance.* Deadweight anchors derive their lateral load capacity by developing friction between their bottom surface and the seafloor (Fig. 31). The maximum horizontal resistance, $R_L$, between the seafloor and the deadweight is a function of the gravitational force imposed by the deadweight on the seafloor and of the coefficient of friction between the deadweight and the seafloor. The coefficient of friction varies widely with seafloor material type from 0.1 to 0.5 with the type of seafloor material and the nature of the deadweight anchor bearing surface. For the most part, on the predominantly soft cohesive seafloor sediments, typical values are 0.1 to 0.2.

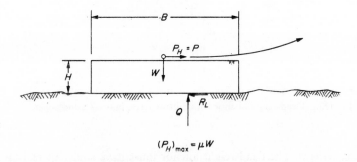

FIG. 31. Simple deadweight anchor with zero line angle, i.e. mooring line horizontal at the anchor.

One purpose of deadweight anchor design is to increase the lateral holding capacity by providing roughened surfaces or teeth to better engage the seafloor surface. Cutting edges, often referred to as skirts in the case of cutting edges around the perimeter of a deadweight, are a common solution (Fig. 32). When the cutting edges have been embedded

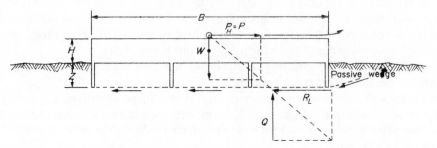

FIG. 32.   Deadweight anchor with cutting edges, zero line angle.

vertically in the seafloor, lateral sliding of the deadweight occurs on shearing zones (possibly planes) emanating from the tip of the cutting edges. Thus the cutting edges serve to force the shear zone into a typically stronger soil. In addition, for the deadweight to move horizontally, the wedge of soil in front of the leading skirt must be pushed up and out of the way providing another component of lateral load resistance.

*Vertical load resistance.* Increasing the mooring line angle at the anchor from zero causes a vertical component of load to be applied to the anchor. This vertical load component must be resisted by a portion of the submerged weight of the deadweight anchor (Fig. 33).

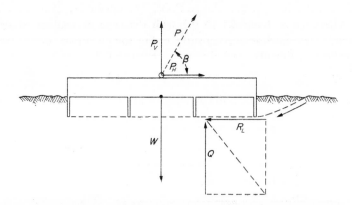

FIG. 33.   Deadweight anchor with cutting edge, non-zero line angle, $\beta$.

*Overturning resistance.* The deadweight anchor must have sufficient resistance to overturning arising from its shape, distribution of mass, and location of mooring line attachment to remain stable under any loading condition. The deadweight must be very flat or squat, that is, it must have a large width, $B$, as compared to its height, $H$. The center of mass must be located as low as possible. Lastly, the mooring line attachment point on the deadweight must be positioned as low on the anchor as possible so as to minimize the lever arm of the horizontal load component, $P_H$. To minimize overturning potential, much of this study assumes the deadweight anchor height to be one-tenth of the width or

$$H = 0.1B \qquad (11)$$

The cutting edge length, $Z$, also influences the overturning potential. Deadweights with low ratios of cutting edge penetration to anchor width, i.e. $Z/B$, will have a lesser potential for overturning.

Deadweight anchors required to resist equal loads from any direction offer little flexibility in varying the point of mooring line attachment. The attachment point must be centrally located allowing only movement to a point lower on the anchor profile. However, deadweight anchors which must resist their heavier loadings from one general direction offer potential for optimizing the attachment point location. Rather than elaborate here, the reader is referred to the Pearl Harbor concrete anchor illustrated in fig. 10(d) of CEL TM-42-76-1 (Valent *et al.*, 1976). This report is limited to consideration of centrally loaded, square-plan deadweight anchors.

*Definition of deadweight anchor failure.* This study assumes that a deadweight anchor has failed if: (1) the bearing capacity of the supporting sediment is exceeded leading to a local or general failure of the deadweight, (2) the lateral load capacity of the sediment is exceeded leading to excessive permanent lateral displacement of the deadweight, and (3) the total "immediate" and "consolidation" settlement is such that the mooring line attachment point to the deadweight is lowered to beneath the pre-existing seafloor surface. (This last mode of failure, that of settlement, has not been treated herein.) Items (1) and (2) separately and individually define any local bearing capacity failure, necessarily involving tilting and some lateral displacement of the deadweight, as a deadweight anchor failure.

Although the assumed definition of anchor failure does require that the deadweight anchor not slide under the ultimate loading condition, the authors allow that this condition may be too stringent. Under certain circumstances it would appear advantageous to design a deadweight anchor to allow some sliding as a mechanism for accommodating peak mooring line loads. Such a sliding system would required sufficient area for movement without adversely affecting the ecosystem or other users of the sea. This study has not treated such a sliding deadweight anchor system.

*Approach.* The sizing of deadweight anchors of the type required for OTEC is dependent primarily on the magnitude of the lateral load component. This study first evaluates the lateral load capacity as a function of deadweight anchor width, $B$, and cutting edge penetration, $Z$. Then the cutting edges are structurally designed to resist the lateral load applied to them with the prime objective of obtaining the cutting edge thicknesses, $t$, required. When cutting edge thicknesses are available, the load required to fully embed the skirts can be calculated. Further, the deadweight configuration can then be optimized to provide the minimum mass required to provide a given lateral load capacity.

Deadweight bearing capacity is treated subsequently to demonstrate that, for a deadweight anchor designed to resist the expected lateral loads for OTEC, bearing capacity failure is not an insurmountable problem.

*Evaluation of lateral load capacity*

*Failure mode.* The lateral load capacity of the deadweight anchors was evaluated assuming the sliding mode pictured in Fig. 33. This failure mode is most desirable because the base shearing zone is forced down into the deepest possible position where the soil shearing strength is greatest for the strength profiles expected. Development of this failure mode is dependent on the ratio of cutting edge spacing to length, $b/Z$; on the net vertical load magnitude at maximum mooring line load; and on the applied overturning moment. The assumption here is that cutting edge spacing, net vertical load, and overturning moment will be adjusted to ensure development of the deep, planar failure zone.

The lateral load capacity was assumed to include resistance from a complete passive wedge failure in front of the leading cutting edge. The existence of a complete wedge assumes no scour about the anchor block. This condition could be expected in most seafloor areas; however, in many category D and some category C environments scour could be a problem. Deadweight anchor design should seek to minimize the potential for scour by streamlining the deadweight and by providing scour protection about the anchor periphery.

In order to minimize the potential for overturning, initial considerations of lateral load capacity assumed a ratio of cutting edge length, $Z$, to deadweight width, $B$, of 0.1. Later evaluation of the lateral load developed as a function of the anchor deadweight required to embed the skirts showed the ratio of $Z/B$ of 0.1 to also be very desirable from the standpoint of design efficiency.

*Analysis of passive wedge component*

1. *Cohesive soil.* The analysis of the passive wedge component for a cohesive soil (categories A, B, and C) utilized the relationship

$$R_p{}^1 = \tfrac{1}{2}\gamma Z^2 + 2sZ \tag{12}$$

where $R_p{}^1 = $ lateral load resistance per unit width of passive wedge (N/m)

$\gamma = $ buoyant unit weight of soil (N/m³)

$Z = $ length of cutting edge (m)

$s = $ undrained shear strength of soil (Pa).

Values of undrained shear strength of the seafloor sediment, $s$, and of the bouyant unit weight of the sediment, $\gamma$, used in the analysis are as presented in Fig. 1 and Table 1. The total lateral resistance arising from the passive wedge is found by multiplying the lateral load resistance per unit width, $R_p{}^1$, by the width of the passive wedge. For the square plan deadweight anchors treated herein, the width of the passive wedge would be the block width, $B$. The passive wedge component of the lateral load resistance is then:

$$R_p = R_p{}^1 B. \tag{13}$$

This value of the passive wedge lateral load resistance is believed realizable at all mooring line directions, i.e. the same value can be realized even when the mooring line is acting along a diagonal of the square deadweight anchor block. When evaluating the lateral load capacity of circular plan deadweight anchors, it would appear reasonable to reduce the average passive pressure acting on a diameter as is done when evaluating the lateral load capacity of cylindrical piles (Czerniak, 1957). Lateral load capacities derived from the passive wedge are presented as a function of anchor side dimension for soil categories A and C in Figs 34 and 35 respectively.

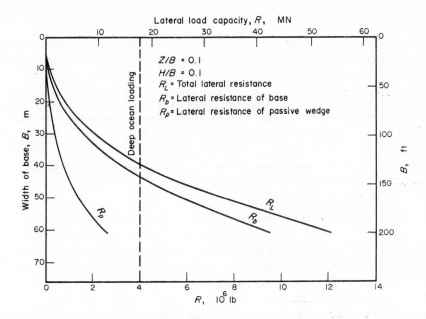

FIG. 34.   Lateral capacity of deadweight anchors vs width in soil category A.

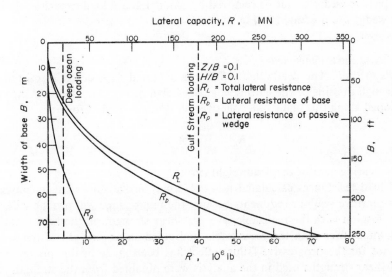

FIG. 35.   Lateral capacity of deadweight anchors vs width in soil category C.

2. *Non-cohesive soil.* The analysis of the passive wedge component for a non-cohesive soil (category D, sand) utilized the relationship

$$R_p{}^1 = \tfrac{1}{2} \gamma Z^2 K_p \qquad (14)$$

where $K_p$ = coefficient of passive lateral earth pressure on the cutting edge.

Table 14 presents values for the coefficient of passive lateral earth pressure on the wall of the leading cutting edge, along with a list of assumptions inherent in the use of these

TABLE 14. COEFFICIENTS OF PASSIVE LATERAL EARTH PRESSURE ON DEADWEIGHT CUTTING EDGE (AFTER TSCHEBOTARIOFF, 1962)

| $\bar{\varphi}$ (deg) | 10 | 12.5 | 15 | 17.5 | 20 | 25 | 30 | 35 | 40 |
|---|---|---|---|---|---|---|---|---|---|
| $\bar{\varphi}$ (rad) | 0.17 | 0.22 | 0.26 | 0.30 | 0.35 | 0.44 | 0.52 | 0.61 | 0.70 |
| $Kp$ | 1.56 | 1.76 | 1.98 | 2.25 | 2.59 | 3.46 | 4.78 | 6.88 | 10.38 |

*Assumptions: a. cutting edge wall is vertical
        b. soil surface is horizontal
        c. soil is non-cohesive, $s = 0$
        d. angle of wall friction, $\delta$, $= \tfrac{1}{2}\bar{\varphi}$.

$K_p$ values. The sand of category D was assumed to have an angle of internal friction, $\varphi$ of 0.52 rad (30°), a conservative value. From Table 14 the $K_p$ value for a $\varphi = 0.52$ rad soil is 4.78. Conservatism has been interjected in the selection of $\varphi = 0.52$ rad because of the unknown cutting edge performance in highly calcareous D soils. The values of the effective friction angle of a calcareous soil on a cutting edge wall, $\delta$, may be considerably reduced due to the soft nature of the carbonate grains relative to steel. The bouyant unit weight of the sediment, $\gamma$, was assumed as 5.5 kN/m³ (35 pcf). The conservative value was used for two reasons: (1) soil in the passive wedge will be less dense than soil under the anchor, and (2) soil in the passive wedge might be removed by scour. Lateral load capacities derived from the pasive wedge are presented in Fig. 36.

*Analysis of base shear component*

1. *Cohesive soil.* The deadweight anchor is assumed to slide on a soil failure plane passing through the tips of the cutting edge as shown in Fig. 33. The resistance to lateral load developed along this plane in a cohesive soil is in its simplest form:

$$R_b = s \times A \qquad (15)$$

where $A$ = bearing area of the deadweight (m²).

Lateral load resistances calculated using (15) may prove somewhat unconservative in some more-sensitive seafloor sediments because the failure zone will probably develop in a progressive fashion with the realizable, average shear strength on the base shearing plane being somewhat lower than the measurable undrained shear strength of the sediment, $s$. No allowance for the progressive failure effect has been made in this preliminary effort. Undrained shear strengths used in the analysis were obtained from the strength profiles of Fig. 1 at soil depths equal to the cutting edge length.

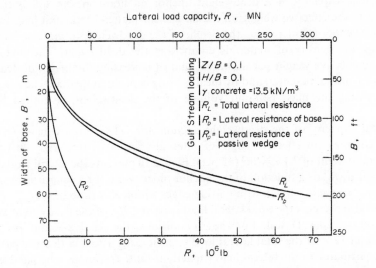

FIG. 36. Lateral capacity of deadweight anchors vs width in soil category D.

Lateral load capacities arising from base shear in category A and C soils have been plotted as a function of anchor width in Figs 34 and 35 respectively. Note that in these figures the ratio of cutting edge length to anchor width, $Z/B$, has been fixed at 0.1.

Figures 34 and 35 also present the total lateral load capacity, $R_L$, the sum of the base shear and passive wedge components, for anchors of square plan. For the $Z/B$ ratio of 0.1 treated and for the category A soil the passive wedge resistance, $R_p$, comprises about 20% of the total lateral load capacity, $R_L$. In the category C soil the passive wedge resistance comprises about 15% of the total lateral load capacity. The slight difference in the contribution of the passive edge is due totally to differences in the shapes of the assumed soil shear strength profiles.

Increasing the $Z/B$ ratio will increase the proportion of ultimate lateral load capacity carried by the passive wedge, $R_p$, because the passive wedge capacity increases as a function of depth squared, $Z^2$, while the base shear capacity increases nearly linearly with depth.

2. *Non-cohesive soil.* The resistance to lateral load developed along the base shear plane in a non-cohesive soil is:

$$R_b = W_{\text{eff}} \times \tan \bar{\varphi} \tag{16}$$

where $W_{\text{eff}}$ = the effective weight or force on the base shear plane (N)

or

$$W_{\text{eff}} = (W + W_s) - P_v \tag{17}$$

where  $W$ = submerged weight of the anchor (N)
$W_s$ = submerged weight of the soil entrained within the cutting edges and above the base shear plane (N)
$P_v$ = vertical force component applied to anchor by mooring line (N).

The lateral load capacity of a deadweight anchor on non-cohesive soil is then directly proportional to the effective weight on the base shear plane. The lateral load capacity can be modified by either (1) increasing the height of the deadweight to increase its mass or by (2) increasing the density of materials comprising the deadweight. The first option for increasing the effective weight per unit area, that of increasing the height of deadweight, is limited by overturning potential and by local bearing capacity potential. The second option, that of increasing the mass per unit volume of the deadweight, is limited by material availability.

Base shear capacity of the deadweight anchors for OTEC on a category D soil were calculated assuming an effective angle of internal friction of 0.52 rad (30°) and a soil submerged unit weight of 7.1 kN/m³ (45 pcf). Figure 36 presents the lateral load component due to base shear, $R_b$, as a function of the anchor block width for a square anchor. The data presented is for an anchor block with cutting edges with a $Z/B$ ratio of 0.1.

Figure 36 also presents the total lateral load capacity, $R_L$, of the deadweight anchor with $Z/B = 0.1$ on category D soil. For these conditions, the passive wedge resistance, $R_p$, comprises about 12% of the total lateral load capacity, $R_L$. The fact that the passive wedge is a minor contributor to the lateral load capacity is quite fortunate. The passive wedge, in some environments, especially in category D soils, could be substantially removed by scour. Because the wedge is a minor contributor, its contribution to the lateral load capacity can be neglected without drastically affecting the anchor block dimensions.

*Results.* The lateral load capacity analysis suggests sizes required for deadweight anchors in various soil categories. These are presented in Table 15. The results of this

TABLE 15.  Length of side of square-plan deadweights required to satisfy lateral load requirements, for $Z/B = 0.1$, line angle $\beta = 0$, $H/B = 0.1$

| Soil category | Lateral load required, $R_L$ | | Length of side, $B$ | |
|---|---|---|---|---|
| | (MN) | (10⁶ lbs) | (m) | (ft) |
| A | 18 | 4 | 40 | 130 |
| B | 18 | 4 | 27 | 90 |
| C | 18 | 4 | 22 | 72 |
| | 180 | 40 | 60 | 197 |
| D | 180 | 40 | 57 | 186 |

phase are significant when considering potential techniques for constructing, transporting and installing deadweight anchors for OTEC. For example, if a deadweight anchor with a lateral load capacity of 18 MN (4 × 10⁶ lbs) were required at a particular OTEC site with category C sediments (Fig. 35), that lateral load capacity could be provided by a square-plan deadweight anchor 22 m (72 ft) on a side. This anchor could be transported in the well of the heavy-lift ship *Glomar Explorer* (well width 22.5 m (74 ft)). However, deadweights of greater width could not be transported within the ship's well, and other transport techniques would be necessary.

*Lateral load capacity as a function of cutting edge length and anchor width*

*Assumptions.* One possible way of increasing the lateral load capacity of an anchor block of given plan dimensions is to increase the length of the cutting edge. Increasing the length and embedment of the cutting edges moves the base shear plane downward into presumably stronger material and increases the size and load capacity of the passive wedge. Unfortunately, the lever arm of the overturning moment is also increased, and thus the lateral sliding failure mode may no longer be controlling.

The review of lateral load capacity as a function of relative cutting edge length, i.e. $Z/B$ ratio, assumes taat lateral sliding would be maintained as the dominant failure mode. This failure mode can be maintained at the expense of restricting the direction of load application to only one direction, i.e. by restricting the multi-direction anchor to a uni-direction anchor. Reduction in the overturning moment is achieved by lowering the mooring line attachment point as illustrated in Fig. 37, or by re-distributing the anchor mass to better counter the overturning moment. The uni-direction anchor is suitable for use in multi-point moorings and in areas of uni-directional loading (e.g. moorings near the axis of the Gulf Stream).

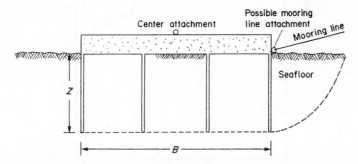

FIG. 37.   Off-center mooring line attachment point, used to decrease overturning moment.

*Results.* Lateral load capacity predictions for deadweights with $Z/B$ ratios other than 0.1 were calculated using the same assumptions and techniques employed above for deadweights with $Z/B = 0.1$ The results are presented in Figs 38, 39 and 40 for soil categories A, C and D respectively.

*Optimal ratios of cutting edge length to anchor width*

*Purpose.* The previous section addressed increasing the lateral load capacity of a deadweight by increasing the cutting edge length. Cutting edge lengths are limited in practice, however, by two factors. One of these factors, aggravation of the overturning problem, was treated earlier. The other, implementation of cutting edge penetration, places efficiency/cost limitations on the cutting edge lengths and section thicknesses. Lengthening the cutting edges increases the lateral load and the effective moment arm acting to bend the cutting edge at its point of attachment to the anchor base (see Fig. 41). To develop the increased bending moment and shear capacity in the cutting edge, the structural section must be increased (thickened). The cutting edge of greater thickness and length requires a greater force, obtained from the deadweight submerged weight, to accomplish complete embedment. The increased costs of the greater structural section in the longer cutting edge and the increased costs for materials, construction, and handling of the heavier deadweight both impose an upper limit on the practical length of cutting edge to be used.

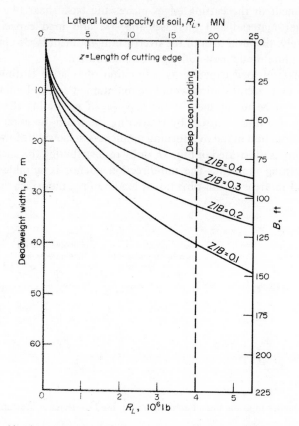

Fig. 38.    Lateral load capacity of deadweight anchor for various $Z/B$ ratios in soil category A.

This report section approaches the subject of optimal $Z/B$ ratios from an idealized stand-point. It assumes that minimum cost is directly related to the minimum deadweight mass required to embed a cutting edge design sufficient to resist the design lateral load. Clarification of this statement is left to the discussion of results for this section.

*Structural design of cutting edges*

1.    *Procedure.* The cutting edges were designed to resist the pressure of a passive wedge as the cutting edge is moved laterally without rotation. The pressure distribution on the cutting edge wall was assumed triangular. Then the force resultant acting on the wall could be taken to operate at the two-thirds point (see Fig. 41). The force per unit width of anchor block $R_p{}^1$ was obtained from the earlier presented passive wedge data. The force times the moment arm yields the bending moment to be resisted at the juncture of cutting edge and deadweight block:

$$M_p{}^1 = R_p{}^1 (2/3) Z. \tag{18}$$

This analysis is conservative, but justified because it is simple. The analysis is conservative because the cutting edges would normally be arranged in a grid pattern beneath the dead-

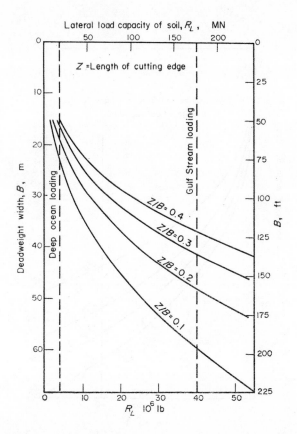

FIG. 39.   Lateral load capacity of deadweight anchor for various $Z/B$ ratios in soil category C.

weight and would interact. Those cutting edges oriented parallel to the mooring line direction would act as shear walls assisting in transferring the mooring load to the perpendicular oriented set of cutting edges and thereby reducing the bending moment in that perpendicular set.

2.   *Results*. Four potential configurations and loading conditions were selected, the bending moments per unit width, $M_p{}^1$, calculated, and then reinforced concrete and structural steel wall sections were designed to resist the applied bending moment. Input parameters and calculated wall thicknesses are contained in Table 16. The design assumed that full loading could be applied to either side of the cutting edge. The reinforced concrete design utilized a working stress design and incorporated 76 mm (3 in.) of cover over the steel. The steel cutting edge design also used a working stress design with efforts to minimize the wall thickness, i.e. the section is not the most economical for resisting the applied moment. The steel section was assumed composed of a row of $H$-pile sections joined into a wall and stabilized by two steel sheets, one on the inside flange, the other on the outside. The design thicknesses, $t$, for the cases analyzed were plotted and joined in Fig. 42 to provide a thickness estimating curve for cutting edges.

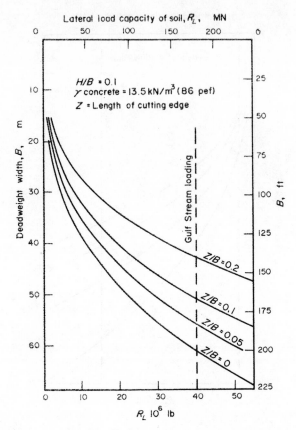

FIG. 40.   Lateral load capacity of deadweight anchor for various $Z/B$ ratios in soil category D.

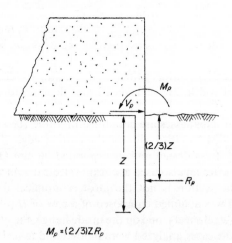

$$M_p = (2/3)ZR_p$$

FIG. 41.   Force-moment diagram for skirt.

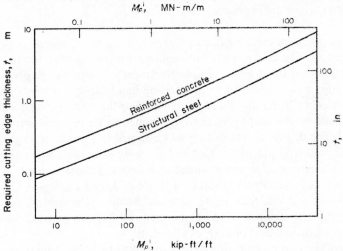

FIG. 42. Estimate of cutting edge thickness required to resist the bending moment at the cutting edge juncture with the anchor block.

TABLE 16. THICKNESS OF CUTTING EDGES FOR SELECTED CASES

| Case | 1 | 2 | 3 | 4 |
|---|---|---|---|---|
| Soil category | A | A | C | C |
| Anchor width, $B$(m) | 43 | 26 | 60 | 37 |
| Cutting edge length, $Z$(m) | 4 | 10 | 6 | 15 |
| $Z/B$ | 0.1 | 0.4 | 0.1 | 0.4 |
| Lateral load on edge per unit width $R_p^1$ (kips/ft) | 5.18 | 35.0 | 31.3 | 131.3 |
| (kN/m) | 75.60 | 510.8 | 456.8 | 1916.2 |
| Maximum bending moment per unit width $M_p^1$ (kips/ft) | 48.700 | 784.50 | 413.00 | 4217.0 |
| (m.MN/m) | 0.217 | 3.49 | 1.84 | 19.0 |
| Cutting edge thickness required, $t$ Concrete (m) | 0.41 | 1.27 | 0.92 | 2.69 |
| Steel (m) | 0.20 | 0.61 | 0.46 | 1.40 |

Please note these thicknesses, $t$, are based on a working stress design while the passive resistances noted are ultimate loadings. Therefore, a change in technique for cutting edge thickness design may be appropriate.

*Load required to penetrate cutting edges*

1. *Assumptions.* The critical factor in dealing with cutting edge penetration is the soil shear strength assumed to be acting. In the treatment of cutting edge penetration, the tip of the cutting edge was assumed to be penetrating in a bearing capacity failure mode with the soil shear strength being mobilized being equal to the undisturbed shear strength (see Fig. 1). The walls of the cutting edge were assumed to be sliding downward in partially remoulded soil. The degree of remoulding was assumed to be at least as great as the minimum value of sensitivity quoted for that soil category (see Table 1 for sensitivities). The penetration of the cutting edge was assumed to reduce the shear strength, $s$, of a category A soil by a factor of 2, a category B soil by a factor of 3, and a category C soil by a factor of 2. The curves of soil side friction, $s_f$, obtained by this technique are compared in Figs 43, 44 and 45 with the recommended values, $\gamma (\bar{\sigma}_m + 2s_m)$, to be used for estimating the soil side friction on driven steel piles (McCelland, 1974).

2. *Analysis.* The cutting edge tips were assumed to be blunt. Penetration of this blunt tip at the expected low emplacement velocities is described by a conventional bearing capacity analysis. This approach is conservative because every attempt will be made to streamline (sharpen) the tips to reduce the tip resistance. However, unknown dynamic

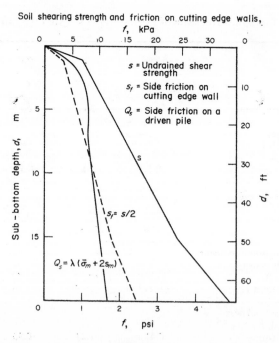

Fig. 43. Comparison of soil friction values for cutting edge penetration analysis in soil category A.

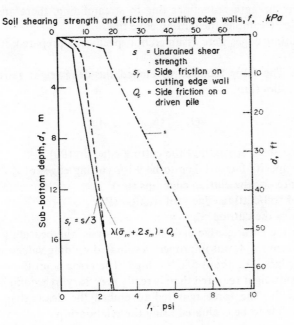

FIG. 44.   Comparison of soil friction values for cutting edge penetration analysis in soil category B.

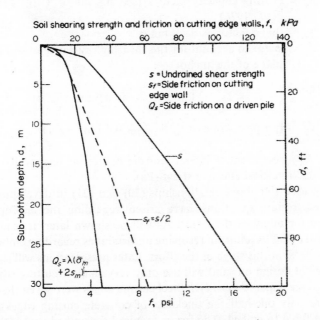

FIG. 45.   Comparison of soil friction values for cutting edge penetration analysis in soil category C.

effects may negate any strength reductions due to streamlining, therefore, no attempt is made in the analysis to streamline the cutting edge tips.

The walls of the cutting edges were assumed not specifically prepared, that is, smoothed or lubricated.

a. *Cohesive soils.* The force required to embed the deadweight skirts, $Q_e$, is to be evaluated using the relationship

$$Q_e = sN_cA + s_fA_s \tag{19}$$

where   $s$ = undrained shear strength at the cutting edge tip (Pa)
   $N_c$ = bearing capacity factor taken to be 9 for cutting edges of $Z/t = 5$
   $A$ = bearing area of the cutting edge tips (m²)
   $s_f$ = mobilized soil/cutting edge wall friction (Pa)
   $A_s$ = side area of the cutting edge (m²).

Unfortunately, during the cutting edge embedment analysis, the bearing capacity term of equation (19), i.e. the term $sN_cA$, was improperly evaluated yielding values of force required to embed the cutting edges, $Q_e$, about 5% too high. The error is on the conservative side indicating a greater deadweight required than is really necessary to accomplish embedment.

b. *Non-cohesive soils.* The force required to embed the deadweight skirts in non-cohesive soils (sand-like) is to be evaluated using the relationship

$$Q_e = A(\tfrac{1}{2}\,\gamma tN\gamma d\gamma + ZN_qs_qd_q) + A_sK\bar{\sigma}_v \tan \varphi \tag{20}$$

where        $t$ = thickness of cutting edge (m)
   $N_\gamma$ and $N_q$ = bearing capacity factors (see fig. 40 of CEL TM-42-76-1, Valent *et al.*, 1976)
   $s_q$ = shape factor = $1 + \tan \varphi$
   $d_q$ = factor to account for the shearing resistance of soil above the bearing plane of the anchor

$$= 1 + 2 \tan \varphi\,(1 - \sin \bar{\varphi})^2 \tan^{-1}(Z/t) \tag{21}$$

$$d_\gamma = (\text{purpose same as } d_q) = 1 + 0.4 \tan^{-1}(Z/t) \tag{22}$$

   $K$ = coefficient of lateral pressure on cutting edge wall
   $\bar{\sigma}_v$ = vertical effective stress (Pa).

After having used the above relationships (20) and (21) in development of the forces required for penetration, questions have arisen regarding the appropriateness of the magnitudes used for many of the terms. As will be shown later, it is not appropriate to spend a great deal of OTEC effort in resolving uncertainties regarding embedment of cutting edges in sand, at least at this stage of the effort. Later presentations will show that the most efficient deadweight design on sand will use only very shallow cutting edges.

3. *Results, force required to embed.* Steel cutting edges were given the greater attention because they are only one-half the thickness of concrete cutting edges and thus require significantly less force to embed. The forces required for embedment, $Q_e$, were calculated for various sizes of square blocks (widths = $B$) and for $Z/B$ ratios of 0.1, 0.2, 0.3 and 0.4 on

cohesive soils and 0.05, 0.10, 0.15 and 0.20 on non-cohesive soils. Intermediate cutting edges were placed beneath the anchor blocks to ensure development of a planar failure plane beneath the anchor. The ratio of cutting edge spacing to length, $b/Z$, varied with the soil type: ratios of $b/Z = 2$, 3, 5 and 6 were used in soil categories A, B, C and D respectively.

Figures 46, 47, 48 and 49 present the forces required to embed cutting edges in soil categories A, B, C and D respectively. The forces for embedment are represented as solid lines. The dashed curve on these plots will be explained in the subsequent section.

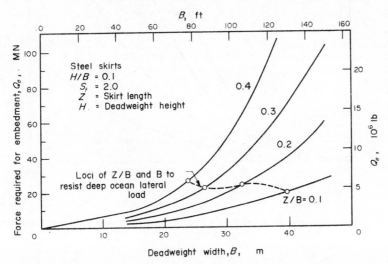

FIG. 46. Required embedment force as a function of deadweight width and height in soil category A.

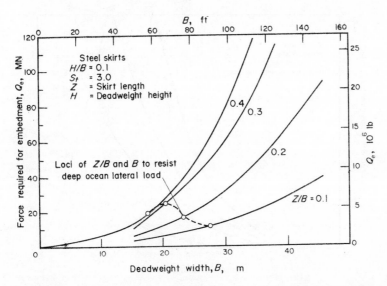

FIG. 47. Required embedment force as a function of deadweight width and height in soil category B.

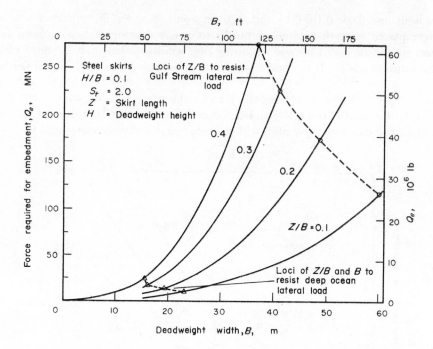

Fig. 48. Required embedment force as a function of deadweight width and height in soil category C.

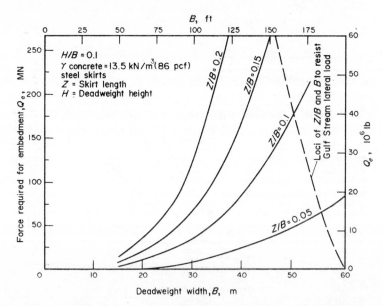

Fig. 49. Required embedment force as a function of deadweight width and height in soil category D.

*Optimum cutting edge ratios*

1. *Approach.* After the development of data above showing the submerged weights necessary to embed cutting edges for deadweight anchors for varying $B$, $Z/B$, $b/Z$, and soil strength profiles, it was then of interest to look for optimum cutting edge characteristics. First, optimum cutting edge characteristics were defined as those that would provide the required lateral load capacity for the minimum required embedment load, $Q_e$. Given this definition, it was then necessary to develop a curve of loci of cutting edge characteristics capable of resisting the required lateral load, i.e. 18 MN or 180 Mn ($4 \times 10^5$ lbs or $40 \times 10^6$ lbs) depending on the environment. The curve of loci is a function of $Z/B$ and $B$ and is represented as a dashed line on Figs 46–49.

The method for developing this curve of loci was to first enter Figs 38, 39 and 40 (and one other unpublished curve for category B soil) and pick out $B$ and $Z/B$ parameters capable of resisting the required loading. With these values for the respective soil category, Figs 46–49 were entered from the $B$ axis until the appropriate $Z/B$ curve was encountered. Thus the loci of cutting edge characteristics capable of resisting the lateral loading were defined.

2. *Results.* Figure 46 indicates for a category A soil there is not much difference in performance for different anchor-block/cutting-edge associations. Please note this evaluation does not touch on the subject of overturning potential which becomes worse with increased $Z/B$ ratio.

Figure 46 indicates that a deadweight weighing between 20 and 25 MN ($4.5 \times 10^6$ to $5.5 \times 10^6$ lbs) will be required to drive a cutting edge system necessary to resist the deep ocean 18 MN lateral load. The waviness of the dashed curve is due to change in the number of intermediate cutting edges in order to maintain a $b/Z < 2$. As indicated before, the cutting edges must be maintained this close together to cause the lateral sliding failure to develop in the horizontal plane of the cutting edge tips.

Figure 47 indicates that the minimum embedment force is required for a cutting edge length to anchor width ratio of 0.1. Embedment force is actually the weight required on the seafloor (for zero line angles). Thus, by minimizing embedment force, weight is minimized and efficiency is maximized. This figure demonstrates that efficiency is maximized at the smaller cutting edge lengths. This trend is very favourable in that overturning should not be a problem with such a wide and squat anchor unit. Figure 47 indicates a required embedment force of 11 MN in order to realize the 18 MN lateral load capacity.

For a category C soil, Fig. 48 indicates again that shorter cutting edges are more efficient, in fact, the optimum $Z/B$ ratio is probably less than 0.1. For the deep ocean environment, a cutting edge embedment force of 11 MN is required to achieve the 18 MN horizontal holding capacity. The subject anchor block would be about 22 m (72 ft) on a side. For the Gulf Stream environment, a cutting edge embedment force of 117 MN ($26 \times 10^6$ lbs) will be required to drive the required skirts to achieve the required 180 MN lateral load capacity. This driving force cannot be placed on the seafloor at one time because the largest lift system known can handle only 62 MN, half of that necessary, and that load only to 1500 m water depth. This limitation in load handling capability suggests that a deadweight anchor in category C soil subjected to a Gulf Stream loading would necessarily:

    a. Incorporate a sophisticated ballasting system to increase the submerged weight of the anchor once it was on the seafloor in order to fully embed the cutting edges.

    b. Incorporate a means for adding mass on top of the anchor to help weight the anchor and drive the cutting edges in a uniform manner.

    c. Incorporate a second anchor system to provide the additional lateral load capacity required, a system such as lateral load resisting anchor piles emplaced through the anchor block.

Figure 49 indicates that for a category D soil (sand) the most efficient section in terms of required submerged weight required for embedding skirts is that with no cutting edges. This phenomenon occurs because the lateral load capacity of a deadweight on a non-cohesive (sand) material is predominantly influenced by the in-water weight of the anchor block. The use of cutting edges to drive the failure zone deeper in the sediment does little to increase the lateral load capacity.

However, cutting edges are desirable, at least around the periphery of the deadweight, to prevent undercutting of the anchor block by scour. Further, the substantial submerged weight must be made available anyway so that weight should also be used to drive some minimal cutting edge design. The base shear design assumes a square plan anchor block of side length $B$ and height $H$ equal $0.1 \times B$ with the block material having an average submerged unit weight, $\gamma$, of 13.5 kN/m³ (86 pcf).

Figure 49 indicates that a deadweight block with a side length, $B$, of 56 m (184 ft), a $Z/B$ ratio of 0.05, and other parameters as above would resist the 180 MN lateral loading of the Gulf Stream. The force required to embed the cutting edges is 64 MN (14.5 × 10⁶ lbs), for all practical purposes the same as the 62 MN load capability of the Glomar Explorer. In practice the 64 MN, 56 m square deadweight with 2.8 m long cutting edges would consist of a structural prestressed concrete compartmentalized box placed on the bottom by a heavy lift system. The cutting edges of this box would be embedded uniformly by the submerged weight of the empty box. Given the box weight of 64 MN and the prestressed concrete unit weight in seawater of 13.5 kN/m³ (86 pcf), the volume occupied by the prestressed concrete would be 4,800 m³ (169,000 ft³). The 56 m × 56 m × 5.6 m high box would enclose a volume of 15,300 m³ (623,000 ft³) of which 70% would be void space at the time of embedment. This void space would then be filled with a material selected to provide the necessary submerged weight (1) to maintain the deep, planar zone of sliding, (2) to resist the overturning moment, and (3) to resist some small uplift load from a non-zero mooring line angle. If the box were entirely filled with concrete, its submerged weight would be 240 MN (54 × 10⁶ lbs), sufficient to resist overturning and some small amount of uplift component.

*Summary.* The lateral load and skirt embedment analyses together then have resolved the general shaping and sizing of the deadweight anchor system for OTEC. Anchors on cohesive sediments will require cutting edge length, $Z$, of $0.1 B$; those on non-cohesive sediments require shorter cutting edges to facilitate embedment with $Z = 0.05 B$ appropriate. Emplacement of anchor sizes sufficient to resist the deep ocean lateral loading of 18 MN (with near zero mooring line angle) is feasible with available technology and equipment. Emplacement of anchors sufficient to resist the Gulf Stream lateral loading of 180 MN (with near zero mooring line angle) on a category D soil (sand-like) is also possible today. Only the anchor conceptual design for resisting the Gulf Stream loading on a category C soil (clay-like) remains a significant problem because of the difficulty of properly applying driving force to uniformly embed the cutting edges. Viable concepts for resisting heavy loads on clay have been suggested but remain to be evaluated and compared.

*Bearing capacity analysis*

*Introduction.* The remaining paragraph of this section briefly presents bearing capacity data. In general, the results of this section suggest that practical deadweight anchors for OTEC in most deep ocean environments will not fail in a bearing capacity mode. However, deadweight anchors on non-cohesive sediments (sands) are susceptible to bearing capacity failure. This section will delve into the reasons for this low bearing capacity on sand, will indicate the magnitude of the capacity reductions, and will delineate steps necessary to make the anchors adequate.

*Analysis.* The vertical load capacity of a very large, shallow, horizontal bearing surface is calculated using the following relationships:

$$Q = B^1 L^1 (0.5\gamma \ B^1 N_\gamma s_\gamma d_\gamma i_\gamma k_\gamma + \gamma Z \ (N_q - 1) \ s_q d_q i_q k_q + s N_c s_c d_c i_c k_c) \tag{23}$$

where   $B^1 = B - 2e_1$

$e_1$ = eccentricity of the load resultant parallel to the $B$ dimension ($B$ is the short dimension)

$L^1 = L - 2e_2$.

$e_2$ = eccentricity of the load resultant parallel to the $L$ dimension ($L$ is the long dimension)

$s_q \ s_\gamma \ s_c$ = factors to account for anchor plane shape

$d_q, d_\gamma, d_c$ = factors to account for the shearing resistance of soil above the bearing plane of the anchor

$i_q, i_\gamma, i_c$ = factors to account for inclination of the resistant load

$k_q, k_\gamma, k_c$ = factors to account for compressibility of the supporting soil.

These factors are all discussed and relationships for evaluating them are presented in CEL TM 42-76-1 (Valent *et al.*, 1976). The values assigned to these factors for the bearing capacity analysis of square deadweight anchors are presented in Table 17.

TABLE 17.   VALUES ASSUMED FOR BEARING CAPACITY
FACTORS IN THE OTEC DEADWEIGHT ANALYSIS

| Cohesive soils, categories A, B and C, $\bar{\varphi} = 0$ | |
|---|---|
| $N_c = 5.14$, $N_\gamma = 0$, $N_q = 1.00$ | |
| (Note: Therefore $N_\gamma$ term of bearing capacity equation $= 0$.) | |
| $s_c = 1.2$ | $s_q = 1.00$ |
| $d_c = 1.05$ | $d_q = 1.00$ |
| $i_c = 0.50$ | $i_q = 1.00$ |
| $k_c = 1.00$ | $k_q = 0.60$ |

| Non-cohesive soils, category D | |
|---|---|
| $s = 0$ | $\bar{\varphi} = 0.52$ rad (30°) |
| $\gamma = 6.75$ kN/m³ (43 pcf) | |
| $N_\gamma = 22$ | $N_q = 18$ |
| $s_\gamma = 0.6$ | $s_q = 1.5$ |
| $d_\gamma = 1.0$ | $d_q = 1.03$ |
| For $R_v = 2 \times P_H$ | |
| $i_\gamma = 0.07$ | $i_q = 0.19$ |
| $k_\gamma = 0.6$ | $k_q = 0.6$ |

Some uncertainty surrounds the selection of the inclination and the compressibility factors. Both sets of factors deserve further experimental verification. The inclination factors $i_q$ and $i_\gamma$ are developed from the expressions presented in TM 42-76-1 for a non-cohesive soil as a function of $P_H/R_v$ (Fig. 50). The authors note that the expressions developed are simplified and may not be totally valid for values of $P_H/R_v$, approaching 1.0. The inclination factor $i_c$ presents a problem in that at least three expressions are available yielding widely differing values for the loading condition of interest. Values vary from 0.25 to 0.7. The authors have relied on the expression of Hansen (1970) for the plane strain condition yielding $i_c = 0.5$ for the OTEC anchor limiting conditions. This assumption may be unconservative in view of Meyerhof's (1963) results and does require verification.

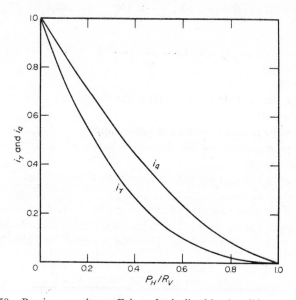

Fig. 50.　Bearing capacity coefficients for inclined loads soil in category D.

The compressibility factors assumed also require verification because of the uncertainty in the assumed soil shear modulus, $G$, for the sediments in question. Reliable estimates of the soil shear modulus, $G$, are required in order to select reliable compressibility factors. Those factors that were selected and listed in Table 17 are best estimates given the available information, but they do require in-depth review.

*Results*

1. *Cohesive*. The bearing capacity analysis indicates that those deadweight anchors on cohesive materials will not fail in a bearing capacity mode. Figure 51 illustrates the relationship of bearing capacity, $Q$, to the lateral load capacity, $R_L$, as a function of anchor width, $B$, on a category A soil. The bearing capacity, $Q$, was calculated assuming the load's resultant applied to the soil was inclined at 0.79 rad (45°). The deadweight was also assumed loaded to ultimate laterally or $P_H$ (load) = $R_L$ (ultimate). Then given the resultant inclination of 0.79 rad, the vertical load component, $R_H$, for the analysis is equal to the lateral load capacity, $R_L$. Thus, the $R_L$ curve in Fig. 51 also represents the vertical load component

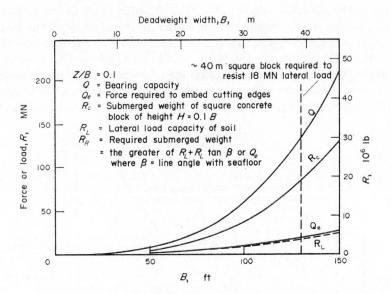

FIG. 51.   Relation of deadweight bearing capacity to various other loadings in soil category A.

$R_v$, applied in the analysis. Obviously, the bearing capacity, $Q$, exceeds the applied vertical load component, $R_v$, by some 750%, thus failure in a bearing capacity mode is very unlikely.

Figure 51 also addresses the driving force necessary to embed the cutting edges, $Q_e$. $Q_e$ is shown to be slightly greater than the vertical load component, $R_v = R_L$, assumed in the bearing capacity analysis. The deadweight anchor would have to weigh the greater amount, $Q_e$, in order to embed the cutting edges. This analysis assumes a zero degree mooring line angle, thus no additional weight is required to balance a vertical component from the mooring line. The submerged weight required for other mooring line angles (assuming $H/B = Z/B = 0.1$) is given by:

$$R_R = R_L + R_L \tan \beta \quad \text{but not less than } Q_e \tag{24}$$

where   $R_R$ = Required submerged weight (N)
        $\beta$ = Mooring line angle with the seafloor (rad).
Note that the bearing capacity plotted in Fig. 51 was based on an assumed submerged weight equal to the applied lateral load. A more precise value of bearing capacity could now be calculated using the required submerged weight ($R_R$) as the vertical load. This was not done since the resulting change in bearing capacity is minimal.

The $Q_e$ values shown in Fig. 51 and elsewhere in this report were calculated assuming that the skirts penetrate evenly and vertically. A canted or inclined deadweight would produce uneven penetration and would result in higher embedment force. Thus, development of means to ensure uniform penetration of the large plan area deadweights required of OTEC may be necessary if uneven penetration is in fact a problem.

The $R_c$ curve represents the submerged weight of a block of prestressed concrete of dimensions $B$ by $B$ square and 0.1 $B$ high, the dimensions assumed for the OTEC deadweight anchors. The potential submerged weight, $R_c$, of a block of these dimensions is 200% greater than the weight required to embed the cutting edges, $Q_e$. Thus, considerable "excess" vertical load carrying potential is available should a non-zero mooring line angle be necessary or should additional weight be required to account for uneven penetration.

On a category C soil (cohesive and strong) the relationship of bearing capacity to the loads being applied is much the same, that is, the applied vertical loads will be considerably lower than the soil bearing capacity (Fig. 52). However, the relationship of the loads $R_L$,

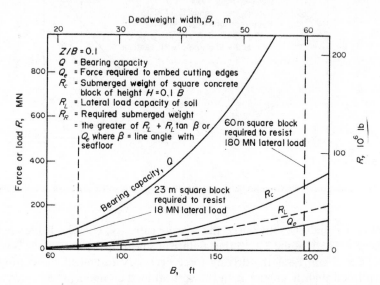

FIG. 52.   Relation of deadweight bearing capacity to various other loadings in soil category C.

$Q_e$, and $R_c$ shifts somewhat. The required submerged weight on the seafloor is again given by Equation (24). As before, the assumed volume of the anchor block, $B$ by $B$ by $0.1 B$ high, provides ample space for weighing material with density equivalent to prestressed concrete.

2. *Non-cohesive.* On non-cohesive sediments, category D soil, the lateral load capacity is influenced much more drastically by load inclination as compared to the inclined load capacity of cohesive soils. Figure 50 shows that for a load inclination of 0.79 rad (45°), the inclination factors $i_\gamma$ and $i_q$ in the bearing capacity Equation (23), are equal to zero. Since the undrained shear strength, $s$, is also zero for non-cohesive soils, the bearing capacity is exceeded. Figure 50 illustrates that the vertical load component, $R_v$, must be increased for any given lateral load, $P_H$, in order to increase the magnitudes of the inclination factors $i_\gamma$ and $i_q$. Figure 53 describes the interdependence of $R_v$, the applied vertical load, and $Q$, the resulting realizable bearing capacity. As the applied load $R_v$ is increased, the bearing capacity, $Q$, is increased, but at a much greater rate. At some point the applied load and the bearing capacity will balance each other. For the case in question, balance occurs at a load component ratio, $P_H/R_v$, of 0.65 corresponding to an applied vertical load, $R_v$, of 280 MN ($62 \times 10^6$ lbs) for an applied lateral load, $P_H$, of 180 MN ($40 \times 10^6$ lbs).

The authors note that a bearing capacity type failure on sand may not be a serious problem. As discussed in the first report, the major result of a bearing capacity type failure is line abrasion on the high side of the block as the load direction shifts. In a uni-directional

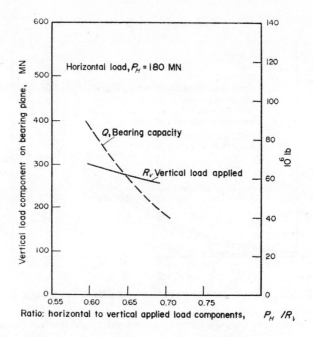

FIG. 53.   Influence of load inclination on bearing capacity, soil category D.

loading environment (Gulf Stream) this line abrasion should not occur. Model testing to be carried out in the next phase of the OTEC effort should clarify the seriousness of a bearing capacity type failure. The extremely large weights indicated by the analysis above may in fact be unecessary if a bearing capacity type failure is found to be of minor consequence.

Figure 54 illustrates the relationship between the various loads involved with a dead-weight anchor on sand as a function of the square block width. The bearing capacity, $Q$, is equal to the required vertical load ($Q = R_R \simeq 1.5\ R_L$). The force $Q_e$, required to embed the cutting edges, of length $Z = 0.05\ B$, is about 25% of the required vertical load, $R_R$, thus no embedment problems are foreseen. In order to achieve the submerged weight required, $R_R$, within the assumed anchor dimensions some special weighting may be necessary. This is so because the weight of a solid, prestressed concrete block, $B$ by $B$ by $0.1\ B$, is less than the block weight, $R_R$, required to achieve a like bearing capacity. The difference in weight between $R_R$ and $R_c$ is relatively small and supplying that weighting, by increased bulk density, should present no problem.

*Findings and conclusions*

1.   The optimum ratio of cutting edge length to square deadweight side dimension, $Z/B$, varies primarily with the soil type:

a.   For cohesive soils, the optimum $Z/B$ ratio is 0.1

b.   For non-cohesive soils, the optimum $Z/B$ ratio is about 0.05 or somewhat less.

2.   Deadweight anchor (square) dimensions necessary to maintain OTEC on station range between 22 and 60 m depending on the loading condition and seafloor material (see Table 15).

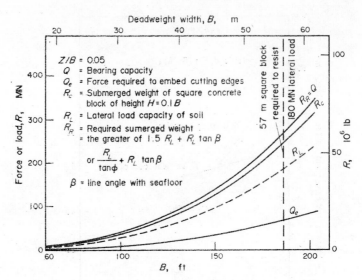

FIG. 54.   Relation of deadweight bearing capacity to various other loadings in soil category D.

3.   In order to prevent excessive tilting of a deadweight anchor (local bearing capacity failure), the effective weight (vertical force component $R_v$) applied to the seafloor must meet certain minimum values:

a.   For cohesive soils, the required weight is given by Equation (24). For the cohesive soils and loads considered, the required embedment force is approximately equal to the applied lateral load.

b.   For non-cohesive soils, the required weight is dictated by bearing capacity, not skirt embedment. To prevent a bearing capacity type failure, a submerged weight of approximately 1.5 times the applied lateral load is necessary. The tremendous weight magnitudes imposed by this requirement may be reduced if testing indicates that a failure in the bearing capacity mode will not lead to catastrophic anchor failure. Also, the coefficients used in the bearing capacity equations need verification for the OTEC loading conditions. Present uncertainties in these coefficients force conservative design. Model tests to reduce this problem, and hopefully to enable deadweight design to be guided by lateral resistance rather than bearing capacity in non-cohesive soils, are included in the next phase of the OTEC effort.

4.   The concept of using a deadweight anchor with cutting edges to moor OTEC platforms on sediment seafloors is viable.

## GROUP CAPACITY

*Factors affecting group capacity*

*Introduction.* Data vailable from which to determine group capacity are extremely limited; thus, maximum use of related data was made. The factors affecting group capacity are not well understood. Depending on the theory or test results one chooses, the following appear to have an effect:

(1) Spacing

(2) Relative embedment depth

(3) Soil characteristics
(4) Anchor type and characteristics
(5) Anchor number and arrangement
(6) Method of installation
(7) Loading geometry (axial, lateral).

The only factor which stands out enough to be considered explicitly is spacing. All theory, model results, and the limited field data available for soft seafloors indicate that decreasing anchor spacing decreases group capacity. Recommended spacings will be discussed and categorized according to anchor type and soil type. Suggested procedures must be considered interim due to the limited data; designs based upon these must use sufficient factors of safety. Once the primary OTEC anchoring choices are defined then the needed data will have to be gathered in order to justify reduced design safety factors.

*Clay seafloor.* After placing anchor flukes, either through dragging, pile driving, augering, ballistic driving, etc., their performance as a function of spacing should be similar. An approach for determining desired spacing in clay was derived from data presented for the prediction of anchor holding capacity by Taylor and Lee, 1972. These data were replotted in Fig. 55 as relative anchor spacing vs soil strengths. Minimum required embedment depth is also indicated by the abscissa. For example, if two anchors of width $B$ in 28 kPa (4 psi) shear strength soil are to exhibit maximum efficiency then they must be placed at least 5 $B$ center to center.

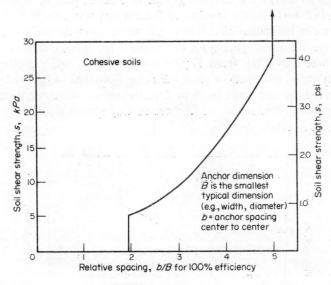

FIG. 55.   Minimum spacing of embedded anchor flukes for cohesive soils to realize 100% group efficiency.

As the anchors are placed closer together, the efficiency of the group is reduced because the individual anchors behave as a single unit. The failure mode change may be of three types. First, if embedded near the soil surface, the failure mode could change from the deep plate mode for a single embedded plate at $Z$; to the shallow plate mode for a group of closely spaced plates all at depth $Z$. A group of three or more plates, embedded to a soil depth near their individual critical embedment depths, would act as one large plate at the same depth $Z$ and the shallow failure mode would occur (see Taylor and Lee, 1972). This failure mode change could reduce the average holding capacity of each anchor in the group to 40% of that of the original, uncombined, originally-spaced anchor capacity.

Second, if the plates are quite deep but still at the same depth $Z$, then when brought close together, their failure zones would interact to reduce their capacity. For instance, if we were dealing with equal square plates and one edge of each were to touch to form a rectangle, then the group holding capacity would be reduced to 92% according to the shape factor.

$$S_p = 0.84 + 0.16 \, B/L. \tag{25}$$

Third, if the plates were stacked one atop the other like the multiple helices of a screw anchor or stacked one behind the other like multiple drag anchors in series on one line, then close spacing of the flukes will result in shearing the soil on the surface of a cylinder or prism encompassing the flukes. This latter mechanism can drastically reduce group efficiency.

The relationship of Fig. 55 might be questioned for individual anchors spaced perpendicular to the direction of pull, because the relationship is based primarily upon the effect of the individual anchors being spaced parallel to the direction of load. However, the results of Langley (1967) show that for groups of bulbous piles in soils of shear strength 24 to 124 kPa (3.5 to 18 psi), a spacing of 5 diameters was sufficient to develop the full pullout resistance of individual piles. These data agree well with the data of Fig. 55.

*Sand seafloor.* Minimum spacing for two anchors for 100% efficiency is plotted in Fig. 56. This data was also taken from Taylor and Lee, 1972, and replotted for simplicity. Typical friction angles of a seafloor sand will range from 0.52 to 0.61 rad (30 to 35°); thus minimum required spacings will be 4 to 5 times anchor width or diameter. The same limitations regarding performance of anchors in clay spaced closer than their minimum holds for anchors in sand.

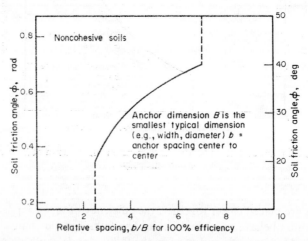

Fig. 56.    Minimum spacing of embedded anchor flukes for noncohesive soils to realize 100% group efficiency.

*Anchor spacing recommendations*

*Conventional (drag burial) anchor spacing.* Conventional anchors can be used singly, or multiply (in tandem or in parallel). The spacings for 100% efficiency in Figs 55 and 56 are recommended. When multiple anchors are used, their group capacity will depend not only upon their spacing and depth of embedment, but also upon how they are connected and installed.

To achieve proper spacing, anchor drag distance for an efficient burial anchor is about six times embedment depth (embedment depth estimated from Fig. 5). For anchors attached in parallel (Fig. 57a) the anchors must be spotted (before setting) a sufficient distance apart to insure that their final spacing (after setting) will be adequate. A somewhat different problem occurs with anchors placed in tandem (see Fig. 57b). If the anchors are not efficient burial anchors, for example, Doris mud anchors (refer to section on conventional anchor extrapolation), then the minimum desired spacing should be increased to prevent the "following" anchors from penetrating the soil disturbed by "prior anchors". The disturbed area could cause a considerable reduction in holding capacity in a sensitive seafloor soil.

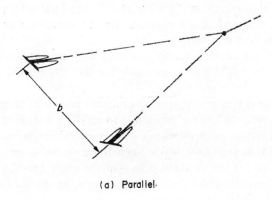

(a) Parallel.

*b* = Anchor spacing for 100% efficiency

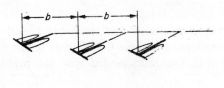

(b) Tandem

FIG. 57.   Tandem and parallel anchor configuration.

*Plate anchor spacing.* Plate anchors should be spaced according to the relationships provided in Figs 55 and 56.

*Pile anchor spacing.* A large amount of data is available on the behaviour of single piles. The difficulty in applying required test loadings has resulted in little performance data for full scale pile groups. This lack of group pile data has resulted in attempts to use single pile data to predict group capacity. One approach has been to define a group efficiency equal to the actual group capacity divided by the sum of individual pile capacities. A typical efficiency formula is shown below. This is known as the Converse–Labarre equation.

$$\eta = 1.00 - 57.3 \frac{D}{b} \times \frac{(n-1)(m) + (m-1)(n)}{90\,mn} \tag{26}$$

where   $\eta$ = group efficiency
$D$ = pile diameter (m)
$b$ = pile spacing (m)
$m$ = number of rows
$n$ = number of columns.

The equation is based on overlapping of bulb stresses around piles which causes a reduction of soil capacity. The Converse-Labarre equation may be used for approximate prediction of friction pile group pullout capacity.

Rough estimates of lateral load group efficiency may be obtained from model tests (Poulous, 1975). No actual field data is available for laterally loaded pile groups in soft seafloor-type materials.

Practical considerations concerning spacing of driven piles were stated by Terzaghi and Peck (1967); their observations appear applicable to uplift or lateral resisting pile groups. They recommend that the distance between pile centers should not be less than $2\frac{1}{2}$ diameters. If spacing is less then the heave of the soil is likely to be excessive and the driving of each new pile may displace or lift the adjacent piles. On the other hand, a spacing of greater than four diameters is uneconomical because it increases the cost of the pile cap without materially benefitting the foundation/anchor.

For intermediate and final design of groups in particular soils a detailed analysis using one of the rational techniques available (Terzaghi and Peck, 1967; Focht and Koch, 1973; Murthy and Shrivastava, 1972) is recommended.

## ANCHORS ON ROCK

*Introduction*

The special anchoring requirements of rock seafloors are discussed in this section. Rock will probably be encountered for some OTEC sites in the Gulf Stream. Thus, the difficulties of anchoring in rock are compounded with the problems imposed by large mooring loads.

There is relatively little experience in rock anchorages, and certainly no experience with the load magnitudes possible with OTEC. Consequently, the analysis below is conceptual. The designs cited are basic ideas which will be refined, altered or combined to achieve optimum performance for a particular situation.

*Environment*

A typical site in the Gulf Stream would have a water depth of about 460 m. It would have a karst-like, possibly cavernous limestone topography. The University of Miami has made several dredge hauls in such areas on the Miami Terrace. Their samples were identified (X-ray crystallography) as amorphous fluor-apatite; a rock in which the carbonate has been replaced by phosphate and fluorine ions. The vertical extent of the rock is unknown. Samples taken were finely laminated. They were composed of alternating fossiliferous and barren layers. Foraminifera were predominant, but not excessive. Deposition was probably Tertiary.

Very little is known about the engineering properties of Gulf Stream rock. This is unfortunate since knowledge of rock strength characteristics is of primary importance in anchor design. Obtaining representative data is a difficult and an expensive task. The engineering properties of marine limestone may vary drastically within a small area (Yang, 1976). For this reason a broad general sampling program would be of little value. Sampling and testing should be done carefully at the actual site location. Actual site topography and rock properties are essential for an optimum anchor design.

*Basic anchor concepts in rock*

The several basic anchor concepts described below are summarized in Table 18. All anchors were designed assuming that horizontal and vertical force components of 180 MN $(40 \times 10^6 \text{ lbs})$ were applied at the anchor.

The seafloor was assumed to be composed of uniform limestone with an average compressive strength of 29 MPa (4200 psi) (Farmer, 1968). A coefficient of friction between concrete and rock of 0.3 was assumed for calculations involving anchor sliding. The bond strength in shear between concrete and a natural rock surface was assumed as 69 kPa (10 psi). A bond strength in shear of 517 kPa (75 psi) between grout and steel and between grout and rock was used for piles and tendons in drilled holes. Several reported designs use similar values (Sakuta *et al.*, 1975; Mutoh, 1975).

*Simple deadweight.* As shown by Table 18, a deadweight of 780 MN $(175 \times 10^6 \text{ lbs})$ is required to resist Gulf Stream loading. The extreme size (83 m $\times$ 83 m $\times$ 8.3 m) of this anchor make handling, transportation and installation major problems.

Two possible techniques for accomplishing these tasks are suggested.

First, the anchor could be fabricated in a large dry dock, then barged or towed to the site using external or internal buoyancy tanks. A buoyancy or heavy lift system would also be required to lower the anchor to the seafloor. Second, the anchor could be cast in place on the seafloor. Again, a heavy lift system would be needed to pre-position the framework on the ocean floor. Technology for pouring concrete in the ocean at a depth of 460 m has not been demonstrated to date, but is probably within reach in the near future.

The advantages of the deadweight are its simplicity and reliability. It does not require that complex underwater operations be carried out in the high energy Gulf Stream environment; and it is not likely to fail catastrophically.

Note that for all deadweight calculations a height to width ratio of 0.1 was used. A ratio of as much as 0.4 is probably acceptable on a rock seafloor. This would reduce the lateral dimension shown in Table 18 to about 0.7 of the stated values.

*Grouted deadweight.* A substantial reduction in anchor size results if grout, injected beneath a deadweight, is used to resist lateral load. Since grout is ineffective in tension, the

TABLE 18. ANCHOR CONCEPTS FOR OTEC IN GULF STREAM,* ROCK SEAFLOOR

| Description | Dimensions (m) | Weight† (MN/10⁶ lb) | Assumptions | Advantages | Disadvantages |
|---|---|---|---|---|---|
| Simple deadweight | 83 × 83 × 8 | 780/175 | Sliding failure‡ | 1. Simple, reliable system | 1. Large weight 2. Bulky 3. Difficult to transport and lower to seafloor |
| Grouted deadweight | 51 × 51 × 6 | 180/40 | Sliding failure§ | 1. Reduced size | 1. Requires site preparation 2. Requires high grout bond quality 3. Difficult to transport and lower to seafloor 4. Requires near-flat, rock seafloor |
| Deadweight in a sinkhole | 57 × 57 × 6 | 255/57 | No lateral sliding | 1. Reduced size 2. Simple installation | 1. Application limited 2. Line abrasion likely |
| Piles (drilled and grouted) Requires 144 0.25-m dia 20 m long steel piles | 20 × 20 × 1 | 13/2.9 | Rock crushing‖ or block uplift¶ failure | 1. Reduced size and weight | 1. Complex installation, requires development of seafloor drilling machine 2. Requires increased on-site construction time 3. High cost |
| Pre-stressed steel wire tendons (drilled, tensioned, grouted) Requires 169 0.06-m dia equivalent, 20 m long steel tendons in drilled holes (0.12 m × 20 m) | 20 × 20 × 1 | 11/2.5 | Sliding or block uplift failure¶ | 1. Reduced size and weight | 1. Complex installation, requires development of seafloor drilling machine 2. Requires increased on-site construction time 3. Requires development of seafloor tendon tensioning equipment 4. High cost |
| Plate anchor in sinkhole | 13 × 13 × 0.7 | 8/1.8 | Fill material and suitable sinkhole available | 1. Simple installation 2. Reduced size and weight | 1. Limited application 2. Line abrasion likely |

*Horizontal and vertical load components = 180 MN (40 × 10⁶ lb).
†Submerged densities: concrete = 13.5 kN/m³ (86 pcf), steel = 67 kN/m³ (426 pcf), rock = 11 kN/m³ (71 pcf).
‡Coefficient of friction concrete-rock = 0.3.
§Bond strength grout-rock = 69 kPa (10 psi).
‖Compressive strength of rock = 19 MPa (4200 psi).
¶Coefficient of friction along vertical rock fracture = 0.5.

deadweight described in Table 18 was designed with a mass sufficient to resist the vertical load. The shear strength of the grout-rock bond is assumed to resist lateral load. A bond shear strength of 69 kPa (10 psi) is required over the entire deadweight bottom surface to resist the 180 MN lateral load. Careful seafloor surface preparation (possibly by suction dredging) is required to insure sufficient bond integrity.

Again, a heavy lift system would be needed to install the anchor. By grouting or gluing the deadweight to the underlying rock, size is reduced, but cost is increased, installation complexity is increased, and a questionable bond with the seafloor exists.

The grouted deadweight concept might be carried one step further. Use of an underwater curing epoxy, instead of a cement grout, could reduce anchor size drastically. Epoxy grouts can sustain tension as well as shear. However, there would be an even more critical dependence on the bond at the seafloor. Experience with underwater epoxies is limited. Their long term performance in a marine environment under heavy loads has not been proven conclusively. Further, bonding of the anchor to a cleaned rock surface may not increase anchor holding capacity since the rock itself may separate along bedding planes. Thus, bonding of an anchor block to a rock surface would be an uncertain operation at best.

*Deadweight—sinkhole combination.* Sinkholes large enough to be explored by submersible occur frequently on the terrace off southeastern Florida. Placing a deadweight into a sinkhole would alter the mechanism for resisting the lateral load component. Much or all of the lateral load component would be transferred directly from the mooring line to the rock at the lip of the sinkhole. The required deadweight size could be decreased accordingly. For example, the anchor in Table 18 was designed to resist a maximum line tension of 180 MN $\times$ 1.41 $=$ 255 MN.

Two major difficulties with utilizing this concept are: (1) there is no guarantee that a suitable sinkhole will be near an acceptable OTEC site, and, (2) line abrasion at the lip of the sinkhole would be severe.

*Driven or drilled piles.* The behavior of piles in rock is not clearly defined. Information dealing with pile systems in rock in the marine environment is limited. The amount of practical installation experience is likewise limited. However, pile systems remain a strong contender for the OTEC anchor in rock because: (1) pile systems are much less dependent on seafloor surface irregularities than a simple or grouted deadweight and (2) piles also substantially reduce the required dimension and material weights as shown in Table 18.

An OTEC anchor comprised of a group of bored and grouted piles in rock could potentially fail in three simple mechanisms: (see Fig. 58).

1. The individual piles will exert high bearing loads against the restraining near-surface rock and could cause crushing of that rock. Crushing of the near-surface rock would allow pile lateral deflection with accompanying increase in bending moment in the pile, and a shifting of restraining load to lower rock layers. The piles could fail in bending or the crushing of rock could progress downward until the piles pulled out.

2. Axial load in a pile could exceed the capacity of the grout to steel bond, the grout to rock bond, or the grout shear strength. Thus, excessive axial load could cause axial pullout of a grouted pile.

3. The rock mass into which the piles are grouted may be weak or fractured. Uplift failure of the entire rock mass could occur. If we assume a weathered, layered limestone, as appears from the limited dredge hauls, along with some jointing, this assumption is likely the case. To reduce the likelihood of such failure, the site survey must identify the probable

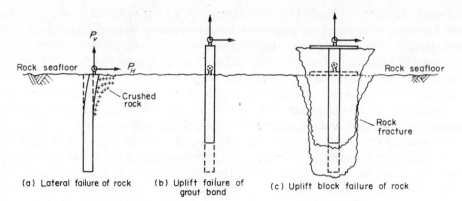

(a) Lateral failure of rock          (b) Uplift failure of grout bond          (c) Uplift block failure of rock

FIG. 58.   Assumed failure modes for pile anchors on a rock seafloor.

rock strength and joint and fault systems, and the pile system must be designed to entrain a sufficient mass of rock to provide the necessary vertical restraint.

The OTEC anchor could also fail in combinations of the above mechanisms depending primarily on the system of discontinuities in the rock. The following analysis considers only the above three simplistic mechanisms.

First, the assumption was made that the entire lateral load on a pile is absorbed by the first 0.3 m of rock adjacent to the pile below the seafloor. Using a conservative limestone compressive strength of 29 MPa (4200 psi), and assuming a 0.25 m diameter pile, lateral resistance per pile was found to be 1.43 MN (320,000 lbs). Roughly 125 such piles are required to resist a lateral load of 180 MN. For simplicity a 12 by 12 pile array (144 piles) was assumed.

Next, the adequacy of the group design against vertical pullout was treated. Failure mode three (Fig. 58c) was found to be more stringent than mode two (Fig. 58b) for the pile spacing and rock condition assumed. Therefore, the pile group in Table 18 was designed to resist uplift failure of the entire block. In other words, pile spacing and length were designed so that the weight of entrained limestone (11 kN/m³) plus fruction developed along the vertical sliding surface balanced uplift force.

Analysis of other cases in the literature suggests that the procedure used is conservative. Even with this conservativeness, a significant anchor size and weight reduction, as compared to a pure deadweight was indicated.

The cost of drilling and grouting the required 144 piles would be high. However, by increasing pile size or perhaps by using a deadweight and battered pile combination, this cost might be reduced. Selection of an optimum design requires a refined analysis procedure based on actual rock properties and loading conditions.

*Pre-stressed tendons.* A similar reduction in anchor size might be achieved by using pre-stressed tendons. First, holes would be drilled through a template into the rock seafloor. High tensile strength steel tendons would then be inserted and grouted in place. Finally, each tendon would be tensioned by jacking down against a template on the seafloor. The template would be forced against the seafloor with a normal force large enough to provide the required frictional resistance to sliding. In effect, the tensioning force acts as an equivalent deadweight force.

For example, using the same coefficient of friction as for the simple deadweight, a tensioning force of 780 MN is required. The total cross-sectional area of steel tendons required is 0.5 m² assuming 250 ksi (1725 MPa) steel. As was the case for piles, the uplift failure mode shown by Fig. 58c governs. Therefore, a template of the same size (20 m), and steel tendons of the same length (20 m) as for piles is needed to entrain the required limestone mass.

Examination of Table 18 shows that little is gained by using post-tensioned tendons rather than piles. The intricacy of installation and resulting high cost would reinforce this conclusion.

*Plate anchor-sinkhole combination.* Another anchor concept which was considered again assumed the availability of a suitable sinkhole near the OTEC site. A large plate is placed into the sinkhole and covered with a fill material. This would remove the major problem of embedment of the large plate size required for OTEC. Approximate plate size was calculated as before and is repeated in Table 18.

This concept faces the same difficulties as noted for a deadweight in a sinkhole, notably limited applicability and line abrasion. In addition, acquiring the necessary fill material would be a formidable task.

## Conclusions and recommendations

Anchor design for a rock seafloor site in the Gulf Stream requires specific knowledge of the local topography as well as rock engineering properties. The investigation revealed that anchoring in such a location will present problems not encountered on normal seafloor sediments. A commensurate increase in anchor cost is probable.

Use of the pile anchor system for the Gulf Stream rock environment is very attractive at present because the pile anchor system can easily accomodate much of the topographic irregularity found in the rock environment and because pile configuration can be altered on site to accommodate variable rock properties. However, pile installation in 460 m of water on an exposed or thinly covered rock seafloor will require the development of some new technology. To resist uplift loads, piles in the rock environment will probably be grouted into bored holes. The equipment for starting such bored holes in thinly covered rock from a floating platform is not available. However, the necessary equipment is not overly complex when compared to present offshore oil-production technology.

Deadweight anchors (simple or grouted) would be an excellent choice for OTEC if: (1) a heavy lift capability is available for placing the anchor and, (2) the ability to pour large concrete masses on the seafloor is demonstrated. Simplicity, economy, and reliability are the primary advantages of deadweight anchors.

## SUMMARY

### Deadweight anchors

Deadweight anchors have been found very suitable for mooring the OTEC platform, especially on unconsolidated sediments. On unconsolidated sediments, the deadweight is fitted with cutting edges about its periphery and at intermediate locations beneath the anchor block. The anchor weight in cohesive soils is dictated by (1) the force required to embed the cutting edges or (2) to prevent overturning, whichever is larger. The cutting edges serve to depress the failure plane in lateral sliding into deeper and stronger soil strata. Through this mechanism, the lateral holding capacity to weight ratio of the deadweight (assumed concrete) is about $\frac{1}{2}$ to 1 (1 to 1 in terms of submerged weight) on cohesive soils. On non-cohesive soils (sands), the lateral holding capacity to weight ratio is about $\frac{1}{3}$ to 1 ($\frac{2}{3}$ to 1 in terms of submerged weight). In this latter case, on sand, the mass required for the deadweight may not be dictated by the weight required to prevent sliding along the base shear plane. Rather, the mass may be dictated by a bearing capacity problem beneath the leading edge of the anchor block: on sands the anchor block is in danger of failing locally in bearing capacity and nosing over into the sediment.

To counter this tendency, the net downward force on the sediment must be increased to increase the effective normal stress across the potential shear zone. Raising the normal stress raises the shear strength of the sand and increases the bearing capacity. In other words, the weight of the block must be increased to prevent a bearing capacity type failure. This means a large increase in the material and installation cost of the anchor. Further work to be done in the next phase of the CEL OTEC effort will determine the consequences of such a failure. Model tests and refinements of the analysis procedure could indicate that a bearing capacity type failure would not adversely affect anchor performance in a unidirectional loading environment (Gulf Stream). Such a result could mean substantial savings in hardware and installation costs.

The use of deadweights on exposed or shallowly buried rock surfaces, such as found on the floor of the Gulf Stream, offers some special problems. The rock surface is relatively un-yielding so it is difficult for cutting edges to bite into it. Nor will the rock surface conform to the deadweight bearing surface. As a result, it is difficult to achieve sufficient lateral load resistance with a deadweight. Effective coefficients of friction on rock are about 0.3, leading to a lateral holding capacity to weight ratio for a deadweight of $\frac{1}{8}$ to 1 ($\frac{1}{4}$ to 1 in terms of submerged weight). On rock, it appears prudent to provide additional techniques for developing lateral load resistance. Short stub piles acting in shear appear excellent candidates for this purpose. Installation of the piles would require some means of developing sufficient thrust on the drill bit in order to start the drill hole, but otherwise should be state-of-the-art.

*Pile anchors*

Pile anchors are very efficient carriers of axial load. Thus, those piles installed vertically as from a drillship can economically resist very high vertical components of mooring line force. However, piles are not efficient in resisting large lateral loads in the almost normally consolidated soil profiles of the deep ocean. Soft surface sediments are not able to provide sufficient lateral load resistance to the deflecting pile. Thus, the load is transferred to lower elevations in the pile, and bending moments are increased. These bending moments dictate structural design of the pile. The weight of hollow cylindrical steel piles is about 20% of the lateral force being resisted (lateral holding capacity to weight ratio of 5 to 1). The steel pile sections must be pressure grouted in the bored holes, especially in the predominantly calcareous ooze sediments. Given the sophisticated construction techniques required, and given the resulting lateral load to weight ratio, pile anchors in typical unconsolidated sea-floor sediments appear of marginal value to the OTEC mooring program. This conclusion presumes that low mooring line angles will prevail for OTEC mooring systems.

Conversely, piles would serve well as OTEC anchors if high mooring line angles were adopted for the OTEC system. There are several reasons why high mooring line angles should be avoided. First, high mooring line angles mean increased mooring line forces: for a mooring line angle of 0.79 rad (45°) the line tension would be 1.4 times the tension for a line angle of 0.0 rad; for 1.0 rad (60°) the line tension would be twice as large. Note that there is not a mooring line available to hold OTEC even at a zero line angle. Raising the line angle increases line tension and compounds the mooring line problem. High line angles also stiffen the mooring which increases dynamic loading at the anchor. In moorings with low line angles, most of the dynamic loading is damped out before reaching the anchor by line motion in the water. Preliminary tests at University of California, Berkeley, suggest that dynamic loadings could result in significant strength reductions in the calcareous ooze sediments that predominate in proposed OTEC siting areas. Thus, dynamic loadings on anchors on calcareous ooze should be minimized to prevent soil strength reductions and anchor pullout. High mooring line angles and resulting high dynamic stresses at the anchor should be avoided.

Pile anchors may serve well to moor OTEC to a rock surface such as exists beneath some parts of the Gulf Stream. The pile sections designed are really steel shear pins which are grouted into rock to provide additional resistance against lateral load.

*Plate anchors*

Plate anchors (direct embedment anchors, screw anchors, etc.) are not suitable for providing the holding capacity required for OTEC in either loading environment. For example, in the deep ocean environment (assuming a zero degree line angle at the seafloor) restraint of OTEC would require a square plate 6.1 m (20 ft) on a side embedded to a soil depth of 30 m (100 ft). To key a 6 m direct embedment fluke approximately 12 m of travel would be necessary. Thus, the initial penetration required of the fluke would be 30 m + 12 m = 42 m (140 ft). A propellant system for embedding these flukes is considered not feasible. Other driving systems, while feasible, would require a complex and very large driving guide on the seafloor and would require considerable equipment development to guarantee the project.

Please note that the above deep ocean loading assumes a horizontal loading. Increasing the mooring line angle would increase the number of 6 m plates required, and would necessitate bridling to equalize the loading. CEL has performed this task in shallow water where assembly could be accomplished subaerially; but installing and bridling flukes remotely, in deep water would be extremely complex and expensive.

*Standard burial anchors*

Drag embedment anchors could be fabricated to provide the required holding capacity for OTEC in the benign, deep ocean environment.

An anchor weighing 64 Mg (140,000 lbs), providing a predicted lateral holding capacity of 18 MN ($4 \times 10^6$ lbs), is feasible. However, drag embedment anchors have two strong disadvantages which render them undesirable for use in the OTEC mooring:

1. Drag embedment anchors must be loaded in only one direction, whereas loading in the deep ocean environment will not be so limited. The direction of load application can be limited by using a multi-point moor with a drag embedment anchor at each leg. This approach would, however, multiply the material cost of the moor by at least four.

2. Drag embedment anchors must be embedded with a zero mooring line angle and, for efficient operation, that line angle must remain near zero. In order to attain a near-zero mooring line angle, a very long scope of line must be used or "sinkers" in the form of concrete blocks of heavy chain must be attached to or integrated into the mooring line to

absorb the vertical component of mooring line tension. Thus, the drag anchor is best used in combination with a deadweight: the deadweight resists the vertical component of load and the drag embedment anchor resists the lateral component of load. Installation of a single leg for a multi-point moor would require precise handling of a heavy deadweight and a large drag anchor simultaneously. This would be an intricate operation in deep water. Other anchor types offer greater versatility and savings in materials. Drag embedment anchors are not desirable for mooring OTEC in the benign deep ocean environment. Drag embedment anchors are even less suitable for use in maintaining the OTEC plant on station in the Gulf Stream environment. To resist possible 180 MN ($40 \times 10^6$ lbs) loads, eight to ten 64 Mg (140,000 lbs) drag embedment anchors, bridled together so as to equally distribute the load, would be required. Proper installation of this drag anchor group, and development and maintenance of the equally distributed loadings, would be very difficult, rendering the concept not suitable for use in the Gulf Stream environment.

## CLOSURE

As a result of this effort, certain constraints on the OTEC mooring design, and indeed possibly on the OTEC plant design, are evident. These constraints have been alluded to here. Coherent presentation of these constraints is made in the final report on the Phase I OTEC Anchor System effort.

*Acknowledgements*—We gratefully acknowledge Mr. Robert Rail for the structural design input; Mr. Joseph Wadsworth for his assistance on the anchoring on rock section; Mr. Lenny Woloszynski for his work on the pile capacity analysis; and Mr. Fred Lehnardt for his excellent efforts in preparing the figures and tables for this document.

## REFERENCES

ANGEMEER, J., CARLSON, E. D. and KLICK, J. J. 1973. Techniques and Results of Offshore Pile Load Testing in Calcareous Soils, in 1973 Offshore Technology Conference Preprints, Houston, Texas, May 1973, Vol. 2, pp. 677–692.

ANGEMEER, J., CARLSON, E. D., STROUD, S. and KURZEME, M. 1975. Pile Load Tests in Calcaereous Soils Conducted in 400 Feet of Water From a Semi-Submersible Exploration Rig, in *Proc. Seventh Ann. Offshore Tech. Conf., Houston*, Vol. 2, pp. 675–670.

ANON. 1968. Department of the Navy, Naval Facilities Engineering Command. NavFac DM-26: Design Manual, Harbor and Coastal Facilities. Washington, D.C., June 1968.

ANON. 1971. Chance Offshore Pipeline Anchoring Systems. Advertising literature OS-15, T. A. Short and Associates, Inc., Houston, Texas, 1971, 7 p.

ANON. 1973. C. G. Doris, Document 1156, Compagnie Generale pour les Developpements Operationnels des Richsees Sous-Marine, Paris, France. Advertising literature, 1973.

ANON. 1976a. American Petroleum Institute. API RP 2A: API Recommended Practice for Planning, Designing, and Constructing Fixed Offshore Platforms. Seventh Edition. Dallas, Texas, January 1976, 47 p.

ANON. 1976b. The Bruce Anchor Mark II, Bruce International Limited, Nassau, Bahamas. Advertising literature, 1976.

BEA, R. G. 1975. Parameters Affecting Axial Capacity of Piles in Clays, in *Proc. Seventh Ann. Offshore Tech. Conf., Houston*, Texas, May 1975, Vol. 2, pp. 611–623.

COLE, M. W. and BECK, R. W. 1969. Small Anchor Tests to Predict Full Scale Holding Power, Preprint, 44th Annual Fall Meeting of the Society of Petroleum Engineers of AIME, Denver, Colorado, Paper Number SPE 2637, 1969, 10 p.

COLP, J. L. and HERBICH, J. B. 1972. Effects of Inclined and Eccentric Load Application on the Breakout Resistance of Objects Embedded in the Seafloor, Texas A & M University, Report No. 153-C.O.E., College Station, Texas, May 1972.

COOMBES, L. P. 1931. Anchors for use on Flying Boats, Reports and Memoranda, Aeronautical Research Committee, May 1931.

CZERNIAK, E. 1957. Resistance to Overturning of Single, Short Piles, Journal of the Structural Division, ASCE, Vol. 83, No. ST2, March 1957, pp. 1–25.

DAVISSON, M. T. and GILL, H. L. 1963. Laterally Loaded Piles in a Layered Soil System, *Soil Mech. Foundat. Div. ASCE*, **89**, 63–94.

DAVISSON, M. T. and SALLEY, J. R. 1968. Lateral Load Tests on Drilled Piers, in Symposium on Deep Foundations, ASTM, San Francisco, California, June 1968.

FARMER, I. W. 1968. *Engineering Properties of Rocks*, p. 57, E. & F. N. Spon Ltd., London, England.

FOCHT, J. A. Jr. and KOCH, K. J. 1973. Rational analysis of the lateral performance of offshore pile groups, in 1973 Offshore Technology Conference Preprints, Houston, Texas, Vol. 2, May 1973, pp. 701–708.

GILL, H. L. and DEMARS, K. R. 1970. Displacement of Laterally Loaded Structures in Non-linearly Responsive Soil, Naval Civil Engineering Laboratory, Technical Report R-670, Port Hueneme, California, April 1970.

HANSEN, J. B. 1970. A Revised and Extended Formula for Bearing Capacity, in Bulletin No. 28, the Danish Geotechnical Institute, Copenhagen, Denmark, 1970, pp. 5–11.

LANGLEY, W. S. 1976. Uplift Resistance of Groups of Bulbous Piles in Clay, M.S. thesis, Nova Scotia Technical College, Halifax, Nova Scotia, Canada, 1967.

MATLOCK, H. and REESE, L. C. 1960. Generalized solutions for laterally loaded piles, *J. Soil Mech. Foundat. Div. ASCE*, **86**, 63–91.

MATLOCK, H. 1970. Correlations for design of laterally loaded piles in soft clay, in 1970 Offshore Technology Conference Preprints, Houston, Texas, May 1970, Vol. 1, pp. 577–594.

McCLELLAND, B. 1974. Design of deep penetration piles for ocean structures, *J. Geotech. Engng Div. ASCE*, **100**, 709–747

MEYERHOF, G. G. 1963. Some recent research on the bearing capacity of foundations, *Can. Geotech. J.* **1**, 16–25.

MEYERHOF, G. G. 1973. Uplift Resistance of Inclined Anchors and Piles, in *Proc. Eighth International Conf. Soil Mech. Foundat. Engng*, Vol. 2, Moscow, U.S.S.R., 1973, pp. 167–172.

MURTHY, V. N. S. and SHRIVASTAVA, S. P. 1972. Analysis of Pile Group Subjected to Vertical and Lateral Loads, in Fourth Annual Offshore Technology Conference Preprints, Houston, Texas, May 1972, Vol. 2, pp. 709–720.

MUTOH, I., IGARASHI, S., SUZUKI, N. and YAMADA, K. 1975. Permanent Anchor for Aquapolis, in Preprints, The 3rd International Ocean Development Conference, Tokyo, Japan, Vol. 2, August 1975, pp. 629–641

POULOS, H. G. 1975. Lateral load-deflection prediction for pile groups, *J. Geotech. Engng Div. ASCE*, **101**, 19–34.

RAECKE, D. A. 1973. Deep-Ocean Pile Emplacement System: Concept Evaluation and Preliminary Design, Naval Civil Engineering Laboratory, Technical Note N-1286, Port Hueneme, California, August 1973, 36 p.

SAKUTA, M., HIRANO, S., MITSUI, M. and MURAKAMI, M. 1975. Field Test of Hollow Type Caisson (The System of V.S.L. Rock Anchor), in Preprints, The 3rd International Ocean Development Conference, Tokyo, Japan, Vol. 2, August 1975, pp. 643–653.

TAYLOR, R. J. and LEE, H. J. 1973. Direct Embedment Anchor Holding Capacity, Naval Civil Engineering Laboratory, Technical Note N-1245, Port Hueneme, California, December 1972, 34 p.

TAYLOR, R. J., JONES, D. and BEARD, R. M. 1975. Handbook for Uplift Resisting Anchors, Civil Engineering Laboratory, NCBC, Port Hueneme, California, September 1975, 153 p.

TERZAGHI, K. and PECK, R. B. 1967. *Soil Mechanics in Engineering Practice*, 2nd edition, John Wiley and Sons, New York.

TSCHEBOTARIOFF, G. P. 1962. Chapter 5, Retaining Structures, in *Foundation Engineering*, edited by G. A. Leonards, p. 462, McGraw Hill Book Co., New York.

VALENT, P. J., TAYLOR, R. J., LEE, H. J. and RAIL, R. D. 1967. State-of-the-Art in High-Capacity, Deep-Water Anchor Systems, Civil Engineering Laboratory, Technical Memorandum M-42-76-1, NCBC, Port Hueneme, California, January 1976, 109 p.

VIJAYVERGIYA, V. N. and FOCHT, J. A., Jr. 1972. A New Way to Predict the Capacity of Piles in Clay, in Fourth Annual Offshore Technology Conference Preprints, Houston, Texas, Vol. 2, 1972, pp. 865–874.

YANG, Z. and HATHEWAY, A. W. 1976. Dynamic response of tropical marine limestone, *J. Geotech. Engng Div. ASCE*, **102**, 123–138.